NOT IN BIP 89/90

The
Fascination of Numbers

The
Fascination of Numbers

by

W. J. REICHMANN

A.C.I.S., F.I.A.I.

ESSENTIAL BOOKS, INC.

Fair Lawn, New Jersey

1957

*First published in 1957
Printed in Great Britain by
Butler & Tanner Ltd, Frome and London*

95378
5-58

Contents

	INTRODUCTION *page*	7
1.	THE MAKING OF NUMBERS	9
2.	SERIES—SHAPES AND SQUARES	17
3.	SERIES (*cont.*)—CUBES AND OTHERS	27
4.	'DIGITAL' ROOTS	38
5.	'PRIMES AND FACTORS'	46
6.	DIVISIBILITY	64
7.	MULTIPLICATION WITH A DIFFERENCE	72
8.	LOGARITHMS AND TRIGONOMETRICAL RATIOS	81
9.	PERFECT NUMBERS AND SOME ODDITIES	87
10.	RECURRING DECIMALS	97
11.	CONGRUENCES	108
12.	IRRATIONALS, IMAGINARY NUMBERS AND CONTINUED FRACTIONS	112
13.	PSEUDO-TELEPATHY	125
14.	FALLACIES	135
15.	MAGIC SQUARES	145
16.	NUMBER MYSTERY	156

Appendices

1.	PASCAL'S TRIANGLE AND THE BINOMIAL THEOREM	161
2.	TRIANGULAR NUMBERS AND COMBINATIONS	165

CONTENTS

3. THE FOUR FOURS PROBLEM—
   EXAMPLES *page* 167
4. 'NAMING THE DAY' 169
5. SERIES AND THE WEIGHTS PROBLEMS 172

INDEX 175

Introduction

It is usual for non-mathematicians to make rapid disclaimers at the mere mention of anything mathematical; yet, at the same time, many will derive great satisfaction from the solving of puzzles which involve numerous mathematical principles.

Everyone must, at some time or other, make use of numbers. A business man may employ a cost accountant to amass and construe volumes of statistics, and a sultan may delegate to his vizier the delicate task of counting his wives; but neither of them can wholly dispense with the processes of counting. Similarly, the shopper who cannot add two figures together on paper is still acutely number-conscious; she will certainly not be satisfied with four bananas if she has paid for five, nor will she be slow to observe and comment upon any discrepancies in her change.

Numbers are, in fact, the concern of everybody in everyday use, and it is for that reason that the present book is intended for the general public rather than for any particular class of reader. It is not in any way designed as a text-book. Instead, it sets out to show, in a simple form, the ways in which numbers, or groups of numbers, are related to each other; how they can be expressed in terms of each other; and the general nature of number behaviour in varying circumstances. It is, in short, an attempt to justify the fascination which numbers exercise over the mind of the student who is prepared to delve into and search out their characteristics. In order to achieve this, a number of detailed explanations of relationships—too often taken for granted—are included.

It is, of course, impossible to divorce one branch of mathematics from other branches. Even if this were possible, however, it would still be undesirable, and far from attempting the impossible, it has been found necessary in certain instances to lay emphasis upon the fact that the various branches are in fact closely allied.

Number theory is the branch of mathematics which, besides being the father of all mathematics, is also that nearest to a non-technical public. This does not mean that the subject is necessarily of an elementary nature. Many of its more advanced problems involve the use of extremely complicated processes of logic, whilst some celebrated problems are still no nearer to any solution whatever. Nevertheless, the greater part of this book is confined to the less abstruse aspects of the subject, and an attempt has been made to render all points under consideration readily intelligible. Where proofs are given or where algebraical notation is employed in any other connexion, the notation has been kept to as simple a form as possible.

In ancient times the philosophers ascribed mystical powers to individual numbers, because of the apparently magical quality observable in their construction and relationships. Although these beliefs may have been swept away over the centuries, the relationships upon which they were based still exist and they well repay a little investigation.

I
The Making of Numbers

There is very good reason to believe that primitive peoples were unable to distinguish any number greater than two. This may seem difficult to accept in the present age, for the science of mathematics has made such rapid and tremendous strides in many directions within comparatively recent times that mere counting has now become, within limits, literally child's play.

Nevertheless it is a fact that the vocabularies of many of these early races contained only 'one', 'two' and 'many' as their counting words. To such peoples the visualizing of any number greater than two was as difficult as it still is for us properly to visualize numbers like a million billion. We may talk about immense numbers with gay abandon, but their size is so great as to make the imagination boggle at any attempt to understand their full significance.

Fortunately few people ever need to make such an imaginative attempt; yet comparatively large numbers are apt to intrude upon the average reader's own interests without his becoming aware of the intrusion. It would be interesting, for instance, to ascertain just how many sportsmen who complete their football coupons week by week do so with the realization that there are more than 536 million different ways of selecting eight teams out of a possible fifty.

Early counting methods undoubtedly depended upon the use of the ten fingers (hence the derivation of the word 'digit'), so that it has been customary for most counting to be effected in groups of ten. Various other systems however were evolved from varying sources and some still survive. The system based upon groups of twenty (instead of groups of ten) presumably required the use of the calculator's toes as well as of his fingers, and we may consider ourselves fortunate that this physical process is no longer required of us!

The Roman system of numbering suggests that at one time counting was effected on one hand only, for this system divided numbers into groups of five. If we study the first five Roman numbers, it is not difficult to relate them to pictures of the human hand. If the right hand is held face upwards one can easily visualize each of the fingers as a vertical stroke of a Roman number, whilst the angle at which the thumb departs from the vertical provides the form of the letter V, the Roman symbol for the number 5.

But whatever the size of the actual groups used, it is this fundamental system of grouping which enables us to deal adequately with the larger numbers. Without the aid of the grouping system each different number would require an entirely different figurate name instead of, as at present, having a composite name derived from the groups of which it is composed. In other words, a totally different symbol would be required for each different number.

Every number is made up from a group of digits and it is the relative positioning of those digits—each to the other—which gives the identity of that number. Just as the order of musical notes decides the nature of a piece of music, so does the order and placing of digits decide the nature of a number. If the order of the digits is changed then an entirely different number will emerge. This is because a digit has a different meaning in different positions—a fact which is essential in a system which uses only ten symbols and yet succeeds in using those symbols to express any conceivable number. Each number is built up to a definite pattern so that a sequence of digits in one particular order can represent only one particular number.

The number 1234, for instance, is made up of one thousand, two hundreds, three tens and four ones, and is really a shorthand form of the expression:

$$(1 \times 1000) + (2 \times 100) + (3 \times 10) + (4 \times 1)$$
$$= (1 \times 10^3) + (2 \times 10^2) + (3 \times 10) + (4 \times 1)$$

If we replace the digits 1, 2, 3 and 4 by the letters a, b, c and d respectively, then the number '$abcd$' is really a form of:

$$1000a + 100b + 10c + d$$

The fact that all numbers have a similar structure is of paramount importance in all that follows. It is not, however, essential that counting should be done in groups of ten as shown above, but the use of such groups was found to be convenient and has therefore become an established practice. Our numbers are said to be in the scale of ten, but these same numbers can be converted to quite different appearances in other scales.

In each scale of notation the basic idea is the same. For example, if the scale we use is x, then we count in groups of x, x^2, x^3 and so on (instead of 10, 10^2, 10^3). Thus in the scale of 7, the number 1234 represents:

$$(1 \times 7^3) + (2 \times 7^2) + (3 \times 7) + (4 \times 1)$$

and is therefore equivalent to the number 466 in the scale of 10. It is important, however, that it should be understood that the number 1234 (in the scale of 7) and the number 466 (in the scale of 10) both represent the same thing. They are merely translations of each other in different languages.

Each different scale has one advantage and one disadvantage as against other scales. The lower the scale, so the fewer *different* symbols are required. In the scale of 7, for example, the symbols 8 and 9 would never be used. The number 8 equals $(1 \times 7) + 1$ and would therefore be 11 in the scale of 7. But the advantage of having less different symbols to bother about is counter-balanced by the fact that, in the lower scales, numbers require a greater quantity of digits for their expression. The number 466 (in the scale of 10) is equivalent to 122021 (in the scale of 3) so that, in the latter scale, six digits are required to represent a number which, in the former scale, required only three digits.

The scale of 10 has been found to work as a good compromise between these twin advantages and disadvantages, but it should be mentioned that some mathematicians advocate the adoption of a scale of 12. The adoption of this scale would, of course, call for the invention of an entirely new symbol for the numbers 10 and 11, since each of these would have to be expressed as one digit instead of as two.

It is a simple matter to convert a number from the scale of 10 to any other scale. The procedure is to divide successively by the new scale number, and the various remainders, taken in reverse order, will give the number in the new scale.

To reduce 466 to the scale of 7, we proceed as follows:

```
7 | 466
7 |  66   remainder 4
7 |   9       ,,    3
        1       ,,    2
```

and this gives the number (in the scale of 7) of 1234.

For a long time in the development of numbering, there was no symbol for nought, the accepted principle being that, as it was impossible to count nothing, there was no point in having a symbol. Nought was not considered a number at all. This view is, of course, no longer tenable if we accept negative numbers as being numbers in the same sense as we accept positive numbers, since nought has its own definite position between -1 and $+1$.

Quite apart from the consideration of nought as a separate number by itself, the introduction of a symbol for nought was of the utmost importance in the construction of the larger numbers, as will be seen when one tries to express the number

THE MAKING OF NUMBERS

10203. Here the nought symbol (or cipher) is used to fill the gaps where no other digits are required and so to prevent the other digits from running into each other.

If, however, we accept nought as a number in its own right, we have to make allowances for certain peculiarities observable in its relation to other numbers.

The division of one number by another can be resolved giving a distinct answer which, for each particular division, is invariable. It is impossible, however, to divide any other number by nought and to express the result in any definite form. The result of such a division is said to be infinity and, in this connexion, it really means that the answer is as large as we like to make it.

The remarkable point about this is that we can use the smallest number imaginable as a divisor and still obtain a definite result, no matter how small the number.

Thus, in the equation $\frac{x}{y} = Z$, where Z is the result of dividing x by y, Z is infinity if y equals nought, and it is clear that we can never express the meaning of infinity by the use of digits. Yet, if Y is not equal to nought, then Z has a definite meaning in terms of digits, and this is so no matter whether the value of Y be positive or negative. As we make Y smaller and smaller (that is, as we approach nought), so Z becomes larger and larger (as we approach infinity).

But we can make Y progressively smaller and yet never reach nought at all. Whatever the original value of Y, we can obviously divide it by 2 and so obtain a smaller number which can never be equal to nought, even though we may approach it ever more closely. Like a desert mirage, the nearer we get to it, so the further off it seems.

On the other hand, the division of nought by a non-zero number has the peculiarity that the answer is always nought.

Similarly, multiplication by nought always gives the answer nought irrespective of what numbers are multiplied. There is nothing really remarkable about this; for if we are short of water, it makes no difference how many buckets we possess if they are all empty. There is nothing in any one bucket and there is nothing in all of them taken collectively.

For a long time negative numbers were suspect and rejected as having no relation to physical life. If numbers are used to represent tangible objects such as, for instance, strawberries, it might be argued that you cannot have a negative number if you cannot also have a negative strawberry. One might as well ask how one is to eat a negative strawberry, but if the analogy is not too strained, it may be said that a negative strawberry is one which has been taken by somebody else and which has probably already been eaten.

The relationship between positive and negative numbers may be more easily understood if they are considered in terms of increments of addition and deduction respectively. As a business firm prospers, so its capital increases by the addition of profits; that is, by the addition of positive numbers to the number representing the size of the capital. When losses are incurred, however, the capital number is reduced by the addition of negative numbers. It may be argued that this is only another way of taking away the positive numbers we have previously added, but what is the position when the capital itself becomes a negative quantity?

For example, where the capital has reached the depths of minus £1000 (meaning in essence that this amount is owed to other people) how, if a further loss accrues, is one to deduct a positive quantity from this amount? It is far more logical to add a negative quantity and the final result is clearly the same.

Just as the realm of numbers may be separated into two divisions differentiating between positive and negative numbers, so may each of these divisions be subdivided into the categories of integral numbers and fractional numbers. An integral number is a synonym for a whole number, whereas a fractional number, as its name implies, is one expressed as a fraction.

We thus have four distinct categories of what are known as

rational numbers—a rational number being one which can be expressed by a finite number of digits—and in each category there is an infinitude of numbers. There is no limit to them at all. So far as positive integral numbers are concerned there is a first number—one—but there is no such thing as a last number, since we can add one to any number and so obtain a larger number. If we take into consideration the fractional numbers as well, then we are faced by the proposition that there is not even a first number for, as we have already shown, there is an infinitude of numbers between 0 and 1. And in the same way as the positive numbers stretch interminably in one direction, so do the negative numbers extend in the opposite direction.

In addition to the rational numbers there are also irrational numbers or 'inexpressibles' as they have been called for the simple reason that they cannot be expressed in normal numerical language. A rational fractional number can be expressed in decimal form either by a finite number of digits or by recurring cycles of digits which always repeat in the same order, but expressions for irrational numbers have no ending and the decimals never recur.

The square root of 2, for instance, can only be calculated to the nearest significant figure, according to the degree of accuracy required; it can never be calculated exactly. In the same way, there is no exact value for π, the symbol used to show, amongst other things, the relationship between the diameter of a circle and its circumference. For practical purposes, this is not a serious complication since the approximate values satisfy most requirements. From the theoretical standpoint, the existence of such numbers opens up wide fields of study which are well beyond the scope of this work, but their nature and basic properties are discussed in a later chapter.

DECIMALS

The behaviour of decimal expressions when multiplied is a source of never-ending mystification to readers who have not fully grasped the fact that a decimal number is only another representation of a fractional number. Confusion arises, for example, because the multiplication of decimal numbers always results in an answer less than either of them, whereas the multiplication of integral numbers always results in an answer greater than either of them (except where the multiplier is negative).

Thus: $5 \times 5 = 25$

But $\cdot 5 \times \cdot 5 = \cdot 25$

That this must be so is clearer when it is appreciated that $\cdot 5 \times \cdot 5$ is the same as $\frac{1}{2} \times \frac{1}{2}$ and is therefore equal to $\frac{1}{4}$. For although the expression $\frac{1}{2} \times \frac{1}{2}$ means 'multiply a half by a half', it also means 'calculate a half of a half' and this must obviously be less than a half.

An extension of the same principle also explains the superficially remarkable behaviour of two numbers, one slightly larger than 1 and the other slightly less than 1, when they are successively raised to higher powers of themselves. The numbers 1·1 and ·9, for example, both approximate closely to the number 1, and it might seem, on first considerations that $(1 \cdot 1)^n$ should always be relatively close in value to $(\cdot 9)^n$. There is, for instance, only a small difference between the values of $(1 \cdot 1)^2$ and $(\cdot 9)^2$, the values being 1·21 and ·81 respectively; but the difference between $(1 \cdot 1)^{100}$ and $(\cdot 9)^{100}$ is immense, the former number being nearly 14,000, while the latter is less than $\dfrac{1}{10,000}$.

Thus, as 1·1 is progressively raised to higher powers it becomes increasingly larger whilst ·9, under the same treatment, becomes progressively smaller.

2
Series—Shapes and Squares

Apart from the common simple structure of numbers resulting from the formation of groups, there are some special numbers which can be built up into 'shapes', such as triangular numbers, squares and cubes. These shape numbers are formed from specific series of other numbers and they themselves form other series.

A series is a group of numbers in which each number is related in a certain and invariable way to the number immediately preceding it and the one immediately following it, and through these two, to all other numbers in the series. The two main types of series are called Arithmetical Progressions and Geometrical Progressions.

An Arithmetical Progression is one in which the difference between any two terms is identical and constant. The numbers 2, 4, 6, 8, —, for example, form such a progression because their common difference between any two consecutive numbers has the constant value of 2.

Numbers may also be in series even if they have no constant common difference, provided that their differences themselves form an Arithmetical Progression. This is usually shown by writing the numbers in one row, then showing their differences in the next row so that these can clearly be seen to form a progression with a common difference, thus:

```
    2       5      11      20      32
        3      6       9      12
           3      3       3
```

A Geometrical Progression differs from an arithmetical progression in that each of its terms is related to the term immediately next to it by a fixed ratio instead of by a common difference. For instance, the numbers 3, 9, 27, 81 ... form a Geometrical Progression, each term being three times as great as the preceding term.

It is possible to generate a series simply by addition or multiplication in conformity with the above principles. The series which are so closely related to the shape numbers may, however, be said to exist in their own right, giving valuable clues to the structure of various types of numbers.

The first such series of importance is that of the triangular numbers which form the progression 1, 3, 6, 10, 15, ... Why these are called triangular numbers is more easily understood if they are represented pictorially thus:

```
   1           3           6          10          15       etc.
                                                   .
                                       .          . .
                           .          . .        . . .
              .           . .        . . .      . . . .
             . .         . . .      . . . .    . . . . .
```

It will be seen that each number is made up from the previous number by adding a further row at the bottom of the design, and that this row of necessity increases by one at each move in order to support the previous number. It follows from this fact that triangular numbers are made up thus:

$$1 = 1$$
$$3 = 1 + 2$$
$$6 = 1 + 2 + 3$$
$$10 = 1 + 2 + 3 + 4$$

and that these are obviously all of a pattern, each being an arithmetical progression carried one stage further than the previous number. Triangular numbers can therefore be represented as addition sums, as on page 19.

From this it is apparent that not only is each triangular number the sum of an Arithmetical Progression, but also that the number of terms in each Progression is the same as the relative position of the triangular number in its own series

```
    1   1   1   1   1   1   1
        2   2   2   2   2   2
            3   3   3   3   3
                4   4   4   4
                    5   5   5
                        6   6
                            7
```

Totals: 1   3   6   10   15   21   28

(i.e. the seventh triangular number is the total of seven terms).

This knowledge makes it possible to calculate any specified triangular number without having to calculate all the intermediate numbers. This is done by using the known formula for the calculation of the totals of Arithmetical Progression $S = \left(\dfrac{a+l}{2}\right)n$, where S is the sum of the progression, a is the first term, l is the last term and n is the number of terms. Where S is a triangular number, we know that the first term is always 1 and that the number which is the last term is always the same as the number of terms.

Thus $S = \dfrac{(a+l)n}{2}$ becomes $\dfrac{(1+n)n}{2}$ or $\dfrac{n+n^2}{2}$

so that if we know n, we also know S. In other words, if we know the position of a specified triangular number we also know what that number is. Thus, for the seventh triangular number, $n=7$ and the required number is $\dfrac{7+7^2}{2} = 28$.

Expressed differently, this relationship can be stated as revealing the fact that half the sum of any number and itself squared is always a triangular number.

Triangular numbers can also be expressed in either of the following ways:

	1	3	6	10	15	21	etc.
(a)	1×1	2×1½	3×2	4×2½	5×3	6×3½	
(b)	2×½	3×1	4×1½	5×2	6×2½	7×3	

that is, as the product of two factors which are themselves members of arithmetical progression. That they can be so expressed is clearly shown by the equation $\dfrac{n(n+1)}{2}$ as in (a) and $(n+1)\dfrac{n}{2}$ as in (b), since these two equations are obviously two aspects of the same relationship $\dfrac{(n+1)(n)}{2}$.

Triangular numbers are closely related to square numbers in at least two ways. Every triangular number is of such a nature that if it is multiplied by 8 and added to 1, the result is always an odd square number. This can be proved as follows. Each triangular number is of the form $\dfrac{n^2+n}{2}$. If this is multiplied by 8 and added to 1, the resultant number is of the form $4n^2+4n+1$. But this can be factorized to $(2n+1)(2n+1)$ or $(2n+1)^2$ and since $2n+1$ must be odd, then the resulting number must be the square of an odd number.

The sum of any two consecutive triangular numbers is always a square number, and all square numbers are formed in this way. This can most easily be represented thus:

$$
\begin{array}{cccccc}
1 & 3 & 6 & 10 & 15 & 21 \text{ etc.} \\
  & 1 & 3 & 6 & 10 & 15 \text{ etc.} \\
\hline
1 & 4 & 9 & 16 & 25 & 36 \;\;,,
\end{array}
$$

There is a simple proof that this is so. It has been shown that the nth triangular number is $\dfrac{n(n+1)}{2}$ or $\dfrac{n^2+n}{2}$. Similarly the $(n-1)$th triangular number is $\dfrac{(n-1)(n-1+1)}{2}$ or $\dfrac{n^2-n}{2}$. Therefore the sum of the nth and $(n-1)$th triangular numbers is

$$\frac{n^2+n}{2}+\frac{n^2-n}{2}=\frac{2n^2}{2}=n^2$$

Thus the sum of the 3rd and 2nd triangular numbers $(6+3)=9=3^2$ and the sum of the 5th and 4th numbers $(15+10)=25=5^2$. This relationship can be shown in another way:

$$\begin{aligned}
1^2 &= 1 & &= (1) \\
2^2 &= 1+3 & &= (1)+(1+2) \\
3^2 &= 3+6 & &= (1+2)+(1+2+3) \\
4^2 &= 6+10 & &= (1+2+3)+(1+2+3+4) \\
5^2 &= 10+15 & &= (1+2+3+4)+(1+2+3+4+5)
\end{aligned}$$

and this shows clearly the nature of the two triangular numbers of which each square number is composed. These expressions can be simplified still further. Thus the expression for 4^2, which is $(1+2+3)+(1+2+3+4)$ can be reduced to $2(1+2+3)+4$. Substituting x for 4, we obtain

$$\begin{aligned}
x^2 &= 2(\text{Sum of 1 to } x-1) + x \\
&= 2\left(\frac{x}{2}\right)(x-1) + x \\
&= (x)(x-1) + x \\
&= x^2 - x + x = x^2
\end{aligned}$$

thus proving the proposition.

Square numbers can also be built up by the addition of Arithmetical Progressions containing all the consecutive odd numbers from 1 upwards, as follows:

$$\begin{aligned}
1+3 &\phantom{{}+5+7+9} &&= 4 &&= 2^2 \\
1+3+5 & &&= 9 &&= 3^2 \\
1+3+5+7 & &&= 16 &&= 4^2 \\
1+3+5+7+9 & &&= 25 &&= 5^2
\end{aligned}$$

It is also observable that the number of terms in each progression is the same as the number whose square results (thus the square of 6 has 6 terms in its progression). Furthermore, the last term in any progression is equal to twice the number to be squared, minus 1. (The last term for $6^2 = 2 \times 6 - 1 = 11$.) Thus all squares may be calculated by the addition of the relevant series of odd numbers, and there are two ways of checking where each series ends.

The foregoing relationship can be proved by the formula for the sum of arithmetical progression $=\frac{1}{2}(a+L)n$.

In the present series, a is always 1; L is always $2n-1$.

Therefore, sum of series $=\frac{1}{2}(a+L)n$
$=\frac{1}{2}(1+2n-1)n$
$=\frac{1}{2}(2n)n=n^2$

The knowledge that all square numbers may be derived by the progressive addition of all odd numbers from 1 upwards, leads to a way of finding values for x, y and z in the equation $x^2+y^2=z^2$.

Examine first $\qquad 3^2+4^2=5^2$
This can be written

$\qquad (1+3+5)+(1+3+5+7)=(1+3+5+7+9)$
or $\quad (1+3+5)+(1+3+5+7)=(1+3+5+7)+9$

The two sides agree because

(a) $(1+3+5+7)$ appears on both sides
(b) $1+3+5=9$

Thus the right-hand number z^2 will always be the sum of two other squares (x^2 and y^2) when the last term in its progression is itself a square, being indeed the square of x; and $y=z-1$.

To take a further example, the next odd square after 9 is 25. Therefore the sum (z^2) of the progression of which 25 is the last term will be the sum of two other squares (x^2 and y^2) one of which (x^2) will be 25, whence $x=5$. z^2 will be $\frac{1}{2}(1+25)13=169=13^2$ and the other square (y^2) will be $(13-1)^2=12^2$.

Thus $\qquad 5^2+12^2=13^2$

Another aspect of the foregoing is the fact that where the sum of two consecutive numbers is a square, the difference between their squares is also a square.

Thus $\qquad 5+4=9$
and $\qquad 5^2-4^2=3^2=9$

Square numbers can also be related to certain others by the Pythagorean formula $a^2+b^2=c^2$ where c is the hypo-

SERIES—SHAPES AND SQUARES

tenuse of a right-angled triangle, and a and b are the other two sides.

Thus (a) $3^2 + 4^2 = 5^2$
      (b) $7^2 + 24^2 = 25^2$
      (c) $11^2 + 60^2 = 61^2$

If we know one of the above figures, it is possible to calculate the other two. In the above examples, taken in turn,

(a) (i) $3^2 = 9$ (ii) $4 + 5 = 9$ (iii) $5 - 4 = 1$
(b) (i) $7^2 = 49$ (ii) $24 + 25 = 49$ (iii) $25 - 24 = 1$
(c) (i) $11^2 = 121$ (ii) $60 + 61 = 121$ (iii) $61 - 60 = 1$

From these we can see that where $a^2 + b^2 = c^2$, then $a^2 = b + c$ and $c - b = 1$. If a^2 is known, a and b can be calculated.

If $a^2 = 121$, then $a = 11$

also $b + c = 121$
and $c - b = 1$

Subtracting: $2b = 120$

and therefore $b = 60$, and $c = 61$.

The above applies only where a is an odd number.

The following is a formula for generating all Pythagorean numbers. If x, y and z be the sides of a right-angled triangle, then for any values of a and b such that $2ab$ is a perfect square, then:

$$x = a + \sqrt{2ab}$$
$$y = b + \sqrt{2ab}$$
$$z = a + b + \sqrt{2ab}$$

Thus if $a = 1$; $b = 8$; then $2ab = 16 = 4^2$.

Then $x = 1 + 4 = 5$
            $y = 8 + 4 = 12$
            $z = 9 + 4 = 13$
and $x^2 + y^2 = z^2$
becomes $5^2 + 12^2 = 13^2 (= 169)$

It should be noted that if $a=2$ and $b=4$ then $2ab$ is still 16 as before, but a different set of figures will emerge. In this case $x^2+y^2=z^2$, becomes $6^2+8^2=10^2$. There is in fact a different set of numbers for each different factorization of $2ab$.

For example, if $2ab=64$, then $ab=32$.

and $\qquad\qquad a=1; \quad b=32$
or $\qquad\qquad a=2; \quad b=16$
or $\qquad\qquad a=4; \quad b=8$

Any of these sets substituted in the formula given will result in a set of Pythagorean numbers.

If $3^2+4^2=5^2$; then it is obvious that $6^2+8^2=10^2$.

For if $\qquad\qquad 3^2+4^2=5^2$
then $\qquad\qquad (2^2)(3^2+4^2)=(2)^2(5)^2$
and $\qquad\qquad (2^2\times 3^2)+(2^2\times 4^2)=2^2\times 5^2$
and $\qquad\qquad (2\times 3)^2+(2\times 4)^2=(2\times 5^2)$
and $\qquad\qquad 6^2+8^2=10^2$

The difference between two consecutive square numbers is equal to the sum of their square roots.

Thus, $\qquad\qquad 72^2-71^2=72+71=143$

This is because, if we follow the algebraical equation $x^2-y^2=(x+y)(x-y)$ we have $72^2-71^2=(72+71)(72-71)$ and since we have chosen two consecutive numbers then $x-y$ (or $72-71$ as here) is always 1, so that this bracket of the factorization can be ignored.

All square numbers are, if even, a multiple of 4 or, if odd, 1 more than a multiple of 4. This is obvious. Every even number is of the form $2n$, and the square of this is $4n^2$ which is plainly a multiple of 4. On the other hand, every odd number is of the form $2n+1$, and the square of this is $4n^2+4n+1$. This is 1 greater than $4n^2+4n$, which is exactly divisible by 4. But if n is odd, then n^2+n is even and $4n^2+4n$ becomes a multiple of 8. It follows that odd squares are really 1 more than a multiple of 8.

The fact that the total of any series of consecutive odd

SERIES—SHAPES AND SQUARES

numbers from 1 upwards represents a square number, suggests that there may be some similar property in a series of even numbers. In fact there are several relationships apparent in such a series.

$$
\begin{aligned}
2 &= 2 &= 2^2 - 2 \\
2+4 &= 6 &= 3^2 - 3 \\
2+4+6 &= 12 &= 4^2 - 4 \\
2+4+6+8 &= 20 &= 5^2 - 5 \\
2+4+6+8+10 &= 30 &= 6^2 - 6
\end{aligned}
$$

It will be seen that the sum of any series of even numbers commencing at 2 and omitting none can be calculated by adding 1 to the number of terms $=(n+1)$, squaring the resulting number $=(n+1)^2$ and taking from this total the same number $(n+1)$ giving a final result of $(n+1)^2-(n+1)$. Each total is thus related to the square of the number which exceeds the number of terms by one.

This leads to another relationship. The expression $(n+1)^2-(n+1)$ is equivalent to $(n+1)(n+1-1)$
$$=(n+1)n=n^2+n.$$

Thus the sum of an even number series also equals n^2+n as follows:

$$
\begin{aligned}
2 &= 2 &= 1^2 + 1 \\
2+4 &= 6 &= 2^2 + 2 \\
2+4+6 &= 12 &= 3^2 + 3
\end{aligned}
$$

This relationship can be proved as follows. The sum of an Arithmetical Progression is $\frac{1}{2}(a+L)n$. In the particular series under consideration, a is always 2 and L is always equivalent to $2n$.

$$\therefore\ S = \tfrac{1}{2}(2+2n)n$$
$$= \frac{2n+2n^2}{2} = n+n^2$$

From this same formula, a further relationship arises, for $n+n^2=(n)(n+1)$ and so the sum of any similar series is equal to the product of the number of terms (n) and the number $(n+1)$.

Thus $$2+4+6=12=3\times 4$$

The totals of even series can thus be expressed in three different ways, as follows:

Series	No. of terms n	n^2+n	$(n+1)^2-(n+1)$	$n(n+1)$	Total of Series
2	1	1^2+1	2^2-2	1×2	2
2+4	2	2^2+2	3^2-3	2×3	6
2+4+6	3	3^2+3	4^2-4	3×4	12
2+4+6+8	4	4^2+4	5^2-5	4×5	20

The numbers appearing in the total column are called oblong numbers and can be shown in a pyramid form:

$$\begin{array}{ccccccccc} 2 & & 6 & & 12 & & 20 & & 30 \\ & 4 & & 6 & & 8 & & 10 & \\ & & 2 & & 2 & & 2 & & \end{array}$$

It will be noticed that the oblong numbers are exactly twice as great as the relative triangular numbers.

Square numbers can also be related to each other in that the sums of the squares of $(n+1)$ consecutive integers—of which the greatest is $2n(n+1)$—is equal to the sum of the squares of the next n integers. Thus, if $n=2$; then $2n(n+1)=12$; and $(n+1)=3$, giving:

$$10^2+11^2+12^2=13^2+14^2$$

Similarly, where $n=3$, we have:

$$21^2+22^2+23^2+24^2=25^2+26^2+27^2$$

3
Series (cont.)—Cubes and Others

It has been shown that all square numbers may be obtained by the summation of arithmetical progressions of consecutive odd numbers from the number 1 upwards. Cube numbers may also be obtained from the same series of consecutive numbers, but in a fundamentally different way. If we write down the series as follows:

1 3 5 7 9 11 13 15 17 19 21 23 25 27 29

the cube numbers may be found as follows:

1^3	Take only the first number in the series		=	1
2^3	,, the *next two* numbers	$(3+5)$	=	8
3^3	,, ,, ,, *three* ,,	$(7+9+11)$	=	27
4^3	,, ,, ,, *four* ,,	$(13+15+17+19)$	=	64
5^3	,, ,, ,, *five* ,,	$(21+23+25+27+29)$	=	125

Thus each cube number is equivalent to the sum of a series, and each series commences where the previous series finishes. This is quite different to the procedure for the extraction of square numbers where each successive series includes all previous series.

It is quite a simple matter to ascertain the first term in each cubic series. Inspection of the above cube numbers shows that the first numbers in each series are as follows:

Series	First number	Derived from
1^3	1	$(1 \times 0)+1$
2^3	3	$(2 \times 1)+1$
3^3	7	$(3 \times 2)+1$
4^3	13	$(4 \times 3)+1$
5^3	21	$(5 \times 4)+1$

The first numbers are in fact all of the form $[n(n-1)+1]$ where the series is n^3, and are themselves in series, thus:

$$\begin{array}{ccccccccc} 1 & & 3 & & 7 & & 13 & & 21 \\ & 2 & & 4 & & 6 & & 8 & \\ & & 2 & & 2 & & 2 & & \end{array}$$

It should also be noted that each cubic series consists of the same number of terms as is represented by its cube root. The series for 5^3, for example, consists of 5 terms.

The full relationships between cube numbers and the series of consecutive odd numbers may be demonstrated as follows. Each progression of odd numbers may be shown in a different way. Thus the series $(7+9+11)$, representing the value of 3^3, is the same as $(7)+(7+2)+(7+4)$ and this, in turn, is the same as $(3\times7)+(2+4)$. This is a constant property, and the series for any cube n^3 may be shown as:

($n \times$ first number in series)
  +(Sum of progression (2, 4, etc.) having $n-1$ terms)

or
$$(n)[n(n-1)+1]+\frac{n-1}{2}[2+(n-1)2]$$

$$=n[n^2-n+1]+(n-1)(1+n-1)$$
$$=n^3-n^2+n\ \ +n^2-n$$
$$=n^3$$

Cube numbers can also be built up as follows; the number of terms in each series again being the same as the number's cube root.

$$1^3 = 1$$
$$2^3 = 1+7$$
$$3^3 = 1+7+19$$
$$4^3 = 1+7+19+37$$
$$5^3 = 1+7+19+37+61$$

The numbers used in these series are called hexagonal numbers because they may be pictorially represented in hexagonal form, and these themselves are related back to triangular numbers to the extent that the $(n+1)$th hexagonal number is obtained by adding unity to six times the

SERIES—CUBES AND OTHERS

nth triangular number. Since we have shown that each triangular number is of the form $\frac{n}{2}(n+1)$, it follows that each hexagonal number is of the form $\frac{6n}{2}(n+1)+1 = 3n^2+3n+1$, and this expression is equal to the difference between two consecutive cubes.

$$(n+1)^3 = n^3 + 3n^2 + 3n + 1$$
and
$$(n+1)^3 - n^3 = 3n^2 + 3n + 1$$

and this proves that the summation of a series of hexagonal numbers will give the relative cube number.

Since all square numbers may be built up from series of odd numbers from 1 upwards, it follows that the addition of any cubes, whose series together use up all consecutive odd numbers rising from 1 upwards, will give a square number; and the root of that square will be equivalent to the sum of the roots of the cube numbers.

Thus, $1+2+3=6$
and $1^3+2^3+3^3 = 6^2 = (1+2+3)^2$
$(1+8+27) = 36$

This is because:

$$
\begin{aligned}
1 &= 1 = 1^3 \\
3+5 &= 8 = 2^3 \\
7+9+11 &= 27 = 3^3 \\
\hline
1+3+5+7+9+11 &= 36 = 6^2
\end{aligned}
$$

This relationship eliminates the separate calculation of each individual cube number in any larger calculation which requires the summation of a series of consecutive cube numbers rising from 1^3. For example, the series $1^3+2^3+3^3+\ldots+25^3$ sums to $(1+2+3+\ldots 25)^2$, which is the square of the sum of an arithmetical progression, and may therefore be shown as

$$\left[\frac{n(n+1)}{2}\right]^2 = \left(\frac{25 \times 26}{2}\right)^2 = 325^2 = 105{,}625$$

Cube numbers can also be represented as the difference between the squares of consecutive triangular numbers.

Since $\quad 1^3+2^3+3^3=(1+2+3)^2$
and $\quad 1^3+2^3\quad\;\;=(1+2)^2$
then $\quad\quad\quad\quad 3^3=(1+2+3)^2-(1+2)^2$
$\quad\quad\quad\quad\quad\;\;=6^2-3^2$

This explains why, for each cube number, only certain consecutive terms of the series of odd numbers are included.

Another structure of the cube numbers is revealed by the fact that each such number is a multiple of 7 or differs from a multiple of 7 by 1. That is, each cube number is of the form $7n$ or $7n\pm1$. Therefore no number can be a cube number if it leaves a remainder of 2, 3, 4 or 5 when divided by 7.

Any 'power' number (square, cube, etc.) can be represented as the sum of a particular arithmetical progression and itself forms part of a series in its own particular power category. There is an interesting relationship between the actual series which form the various 'power' numbers. This can be seen more clearly if the various series are written out and reduced to progressions having a constant difference between terms.

SQUARES

$$\begin{array}{cccccc} 1 & 4 & 9 & 16 & 25 & 36 \\ & 3 \quad 5 \quad 7 \quad 9 \quad 11 & & & & \\ & 2 \quad 2 \quad 2 \quad 2 & & & & \end{array}$$

(Thus after *one* step, the series is reduced to a progression with a common difference of 2.)

CUBES

$$\begin{array}{cccccc} 1 & 8 & 27 & 64 & 125 & 216 \\ & 7 \quad 19 \quad 37 \quad 61 \quad 91 & & & & \\ & 12 \quad 18 \quad 24 \quad 30 & & & & \\ & 6 \quad 6 \quad 6 & & & & \end{array}$$

(After *two* steps, the series is reduced to a progression with a common difference of *6*.)

FOURTH POWERS

1		16		81		256		625		1296
	15		65		175		369		671	
		50		110		194		302		
			60		84		108			
				24		24				

(After *three* steps, the series is reduced to a progression with a common difference of *24*.)

As the power is raised by 1 degree, the number of steps is increased by 1 and the resulting common differences are always related. Thus in the series of square numbers, 1 step is required and the common difference is 2 (or 2×1 as it really is). For cubes, the number of steps is $1+1$ (=2) and the common difference is $3 \times 2 \times 1$ (=6). For fourth powers, the steps are $1+1+1$ (=3) and the common difference is $4 \times 3 \times 2 \times 1$ (=24).

The series for all power numbers are similarly related to each other. Each successively higher power series requires one more step than the previous series before it reduces to a common difference; and the common difference for any one series is found by multiplying the common difference of the next lower series by the same number as is represented by the index of the power number.

This means that in the series of numbers raised to their nth powers, there will be $n-1$ steps before a common difference is revealed, and this common difference will be equivalent to $(n)(n-1)(n-2) \ldots (2)(1)$.

The respective common differences, 2, 6, 24, 120, as well as the common difference just shown for nth power numbers, themselves form the series of factorial numbers which are of great importance in the binomial theorem and the theory of probability. The factorial number n is usually written $n!$ and is equivalent to the product of all the integers from 1 to n inclusive; so that, for example, $4! = 1 \times 2 \times 3 \times 4$.

This enables us to reduce the form of the common difference in the nth power series from $(n)(n-1)(n-2) \ldots (2)(1)$ to the simple expression $n!$

The next regular pattern to appear in 'power' numbers is that revealed by their end-digits—that is, the last digit in each number. If the squares of consecutive integers are written out as follows, there are two points which are immediately apparent.

	100		400	
1	81	121	361	441
4	64	144	324	484
9	49	169	289	529
16	36	196	256	576
25		225		625

These numbers have been placed consecutively in columns which show the numbers downward in the odd columns and upward in the even columns. This makes it easier to check that in each set of ten numbers the end-digits are, in order, 1, 4, 9, 6, 5, 6, 9, 4, 1, 0, and that, moreover, of this set of numbers the digits preceding the digit 5 appear again after 5, but in reverse order.

These end-digits repeat in the same order indefinitely as each new square number is reached, so that it may always be stated with confidence that no number whose end-digit is 2, 3, 7 or 8 can possibly be a square number.

It may be wondered why two numbers whose end-digits are 9 and 1 respectively each give numbers with the end-digit 1 when squared. There is an elementary reason why this should be so and, stated simply, it may be said that it is because both numbers differ from a multiple of 10 by 1, and is based on the fact that $(-1)^2$ gives the same result as $(+1)^2$.

Any number ending in 9 can be expressed as a multiple of 10, less 1 (that is, $10n-1$). Any number ending in 1 can be expressed as a multiple of 10, plus 1 (that is $10n+1$). If these two expressions are squared, the results are:

$$(10n-1)^2 = 100n^2 - 2(10)n + 1$$
$$(10n+1)^2 = 100n^2 + 2(10)n + 1$$

SERIES—CUBES AND OTHERS

Since both $[100n^2-2(10)n]$ and $[100n^2+2(10)n]$ must clearly be multiples of 10, it follows that both squares must be one more than the multiples of 10. This is another way of saying that they will both have 1 as end-digit.

For the same reason, the squares of any two numbers having end-digits which are complementary (i.e. add together to make 10; such as 8 and 2, 7 and 3, etc.) will themselves have identical end-digits.

The same principle can be applied to cube numbers. Any cube ending in 9 can be written $(10n-1)^3$ and this equals $(10n-1)[(10n)^2-2(10)n+1]$. If the brackets are removed all the terms in the final result, except the last term, will be multiples of 10. The last term is $(-1)(+1) = -1$. The expression is therefore a multiple of 10, less 1; and the end-digit is therefore 9.

The end-digits of squares are limited to 0, 1, 4, 5, 6 and 9. Those of cubes, however, use up 0 and all nine digits. Numbers ending in 0, 1, 4, 5, 6 or 9 have cubes with the same end-digits respectively, whereas numbers ending in 2, 3, 7 or 8 have cubes the end-digits of which are complementary to 10. Thus if the end-digit of a number is 2, then its cube will have the end-digit 8.

As fourth powers are the squares of square numbers it follows that the end-digits of fourth powers are limited to the end-digits of the squares of 0, 1, 4, 5, 6 and 9—that is 0, 1, 6 or 5. Fifth powers have exactly the same end-digits as have their roots. This is another way of saying that the difference between any two consecutive fifth power numbers is always a number having 1 as its end-digit.

$$2^5 - 1^5 = 32 - 1 = 31$$
$$6^5 - 5^5 = 7776 - 3125 = 4651$$

This can be proved as follows:

$$(n+1)^5 = n^5 + 5n^4 + 10n^3 + 5n + 1$$
$$(n+1)^5 - n^5 = 5n^4 + 10n^3 + 10n^2 + 5n + 1$$
$$= 5(n^4 + 2n^3 + 2n^2 + n) + 1$$

But $(n^4+2n^3+2n^2+n)$ must be even and can be represented

as being of the form $2x$ where $x=\dfrac{n^4+2n^3+2n^2+n}{2}$. The difference between two consecutive fifth powers is therefore of the form $5(2x)+1=10x+1$, and the resulting number must therefore always have 1 as its end-digit.

A table of end-digits for numbers up to the seventh powers is given below. It will be seen that all fifth powers have the same end-digit as the original number; all sixth powers have the same end-digits as squares and all seventh powers have the same end-digits as cubes. This process is continuous, every power number having the same end-digit as the number four degrees lower.

End-digit of Number	Squares	Cubes	Fourth	Fifth	Sixth	Seventh
1	1	1	1	1	1	1
2	4	8	6	2	4	8
3	9	7	1	3	9	7
4	6	4	6	4	6	4
5	5	5	5	5	5	5
6	6	6	6	6	6	6
7	9	3	1	7	9	3
8	4	2	6	8	4	2
9	1	9	1	9	1	9

We have already given numerous relationships among squares and cubes. The particular behaviour of numbers consisting entirely of nines often points the way to discovering different relationships.

Consider first, the following squares:

$$9^2 = 81$$
$$99^2 = 9801$$
$$999^2 = 998001$$
$$9999^2 = 99980001$$

Any of the above square numbers can be obtained merely by placing a 9 in front of the previous square and adding a further nought between the digits 8 and 1. It will also be noticed that if the square number is divided at the centre into two separate numbers and these are added together,

SERIES—CUBES AND OTHERS

then the original root number will appear (thus, 998+001 =999).

Now the square number 81 can also be obtained from the expression $(8 \times 10) + 1$, and this expression can be put into terms of the root number (9) thus:

$$81 = (9-1)(9+1) + 1$$

Similarly $\quad 9801 = (99-1)(99+1) + 1$

Substituting x^2 for 81 or 9801, we obtain:

$$x^2 = (x-1)(x+1) + 1$$

and we can immediately prove that the square of any number can be obtained by adding 1 to the multiple of the next number below and the next above the original number, for the equation $x^2 = (x-1)(x+1) + 1$ is the same as $x^2 - 1 = (x-1)(x+1)$ and this is true for all values of x.

The cubes of 'all-nine' numbers are also of a pattern.

$$9^3 = 729$$
$$99^3 = 970299$$
$$999^3 = 997002999$$

For each successive cube, we take the previous cube, add another digit 9 at the front and at the end, and add a further nought between the digits 7 and 2. The reason for this is not immediately obvious, but a study of the following elementary cube numbers uncovers the relationship.

$$2^3 = 8$$
$$3^3 = 27$$
$$4^3 = 64$$
$$5^3 = 125$$

These may be written differently:

$$2^3 = 8 = 0 + (3 \times 2) + 2$$
$$3^3 = 27 = 16 + (3 \times 3) + 2$$
$$4^3 = 64 = 50 + (3 \times 4) + 2$$
$$5^3 = 125 = 108 + (3 \times 5) + 2$$

To begin with we may assume that each cube is partly built up from three times its root number, plus 2. The original clue to this assumption was given by the fact that the last two digits of 9^3 ($=729$) are 29 and this equals $(3\times 9)+2$ whilst the last three digits of 99^3 ($=970299$) are 299 and this equals $(3\times 99)+2$. It is then necessary to show how the remaining portion of each cube number is built up (that is, the numbers in the series 0, 16, 50, 108, etc., shown above).

For the cube of 4, this number is 50 and this equals $2(5)^2$. For the cube of 5, the relevant number is 108 and this equals $3(6)^2$. These two examples at once reveal their pattern, and we now have

$$4^3 = 2(5)^2 + (3)(4) + 2$$

giving the general equation:

$$x^3 = (x-2)(x+1)^2 + 3x + 2$$

which may be proved algebraically since $(x-2)(x+1)^2$ equals $x^3 - 3x - 2$.

This is not a useful equation for evaluating cube numbers in general, but it renders the particular cases of 'all-9' cubes very simple indeed.

$$\begin{aligned}999^3 &= (997)(1000)^2 + 3(999) + 2\\ &= 997{,}000{,}000 + 2997 + 2\\ &= 997{,}002{,}999.\end{aligned}$$

It can also be used for certain other selected numbers.

$$\begin{aligned}49^3 &= (47)(50)^2 + 3(49) + 2\\ &= 117{,}500 + 147 + 2\\ &= 117{,}649.\end{aligned}$$

Here it may be mentioned that although the powers of numbers consisting entirely of nines are very much of a pattern, so that one answer can be derived from another merely by the insertion of other digits, this does not result from any magic power possessed by the digit 9. It results instead from the fact that 9 is one less than 10; 99 is one less than 100, and so on.

The difference between any two consecutive cube numbers may be found by a simplified process. The difference between 10^3 and 9^3 is 271 (that is, $1000-729$) and this difference equals $[3 \times 9 \times 10) + 1]$. This gives a formula which applies to any pair of cubes x^3 and y^3, where $x = y+1$. Thus

$$x^3 - y^3 = 3xy + 1$$

This can be proved by substituting $(y+1)$ for x. Then

$$\begin{aligned}(y+1)^3 - y^3 &= y^3 + 3y^2 + 3y + 1 - y^3\\ &= 3y^2 + 3y + 1\\ &= 3y(y+1) + 1\\ &= 3y(x) + 1\\ &= 3xy + 1\end{aligned}$$

It has already been shown that it is possible to find (in various different ways) a square number such that it is equivalent to the sum of two other square numbers. That is, $x^2 = y^2 + z^2$. No similar number of a higher power has ever been found to satisfy the same conditions. Thus, no satisfactory solution exists for $x^n = y^n + z^n$ where the exponent n is greater than 2. The renowned mathematician Fermat claimed to have found a proof why this should be impossible but his claim has never been substantiated. Other mathematicians have proved it to be impossible for many values of n but no one has succeeded in producing a general proof for *all* values of n.

It is, however, possible to find a cube number equal to the sum of *three* other cubes.

$$3^3 + 4^3 + 5^3 = 6^3$$
$$1^3 + 6^3 + 8^3 = 9^3$$

4
'Digital' Roots

In the last chapter it was shown that numbers consisting entirely of nines have properties peculiar to themselves. The digit 9 is, of all the digits, the most persistent, and many other numbers can be reduced to or shown to have an affinity for 9 in a number of ways. For a full appreciation of these facts, it is necessary to understand the process of extracting digital roots.

The digital root of any number consists of only one digit. There are thus only nine different roots possible, but these few roots can be made to perform a lot of work. The digital root of a number is obtained by adding its digits together and treating the resulting number in the same way until, after a certain number of stages, the final number consists of only one digit. This final number is the digital root of the original number.

The digital root of the number 6542807 is therefore obtained thus:

(a) $6+5+4+2+8+0+7=32$
(b) $3+2=5$

Roots can be also calculated as being the remainder when a number is divided by 9, except where that number is exactly divisible by 9. If a number is exactly divisible by 9, then there is no remainder, but it is obvious that no number can have nought as its digital root since the latter is the sum of individual digits. In point of fact, all numbers which are exactly divisible by 9 also have 9 as their digital root (i.e. 18 has $1+8=9$; 378 has $3+7+8=18$, and $1+8=9$).

'DIGITAL' ROOTS

Why this should be so is easier to demonstrate by showing the process as in the following example.

The number 54362 can be expressed as $(9 \times 6040) + 2$. Here, 2 is the remainder and therefore the digital root. The number 54360 can likewise be expressed as $(9 \times 6040) + 0$, whence we might assume the root to be nought. But it can also be expressed as $(9 \times 6039) + 9$, giving 9 as a 'remainder' and also as the digital root. It is interesting to note that 54360 can also be expressed in other ways, as:

$$(9 \times 6038) + 18$$
$$(9 \times 6037) + 27$$
$$(9 \times 6036) + 36$$

In each case however the 'remainders' can themselves be reduced to the digital root 9.

The same principle can be applied to all numbers. The number 54362 can be expressed as:

$$(9 \times 6040) + 2$$
$$(9 \times 6039) + 11$$
$$(9 \times 6038) + 20 \text{ etc.}$$

and in each case the 'remainders' all reduce to the digital root 2.

It will therefore be seen that any number x will have the same digital root as x plus or minus any multiple of 9.

If
$$x = 64 \quad \text{Root} = 1$$
$$x + 9 = 73 \quad ,, \ = 1$$
$$x - 9 = 55 \quad ,, \ = 1$$
$$x - 27 = 37 \quad ,, \ = 1$$

From this simple example it follows that if two numbers (or more) are added together, the root of the total bears a fixed relationship to the roots of the numbers being added.

64	Root = 1	
9	Root = 9	
Add: 73	Add: = 10	Reducing to: Root = 1

Since this relationship is fixed, it follows that if the digital root of the total number cannot be derived from the digital roots of the individual numbers, then the total number is incorrect. In other words, all additions can be checked by checking the digital roots.

Thus,   64321   Root = 7
         13679   ,, = 8
          5032   ,, = 1
   Add: 83032   16   (Reducing to 7)

The digital root of 83032 is 7, and the total of the individual roots also reduces to 7.

Before proceeding further, it is to be noted that this is not an infallible check for additions. Although an addition cannot be correct if the roots are not in the right relationship, it does not follow that an addition is necessarily right because the roots are correct. Because all negroes have curly hair, it does not follow that all people with curly hair are negroes.

In the example:

                5432   Root = 5
                1618   ,, = 7
        Add: 7005   Add: 12
        Final Root 3   3

the roots agree but the addition is incorrect. This is because the digits in the total are correct but in the wrong order. The correct total is 7050 and this obviously has the same root as 7005.

In the same way, the extraction of digital roots will not reveal any error where the difference between the correct and incorrect totals is a multiple of 9, since we have already noted that the number x will always have the same root as x plus or minus 9.

Subject to the same qualifications, subtraction can be checked by deducting roots.

                64321   Root = 7
                12120   ,, = 6
   Subtract: 52201   Subtract: 1

Similarly if two numbers are multiplied together, the root of the resulting number is the same as the product of the roots of the original numbers

$$\begin{array}{rl} 34 & \text{Root} = 7 \\ 21 & \text{,,} = 3 \end{array}$$

Multiply: 714   Multiply: 21

Final Root   3   3

and again, if one number is divided by another the root of the dividend is the same as results from the division of the first root by the second root.

$$\begin{array}{rl} 5921 & \text{Root} = 8 \\ 31 & \text{,,}   4 \end{array}$$

Divide: 191   Divide: 2

Final Root   2   2

But it is possible to have two numbers the larger of which has the smaller digital root. The two numbers 34 and 17 are such a pair, their roots being 7 and 8 respectively. How can 34÷17 be related to the digital process, when the division of the respective roots results in $\frac{7}{8}$ which seemingly bears no relation to the final root of 2? The answer is simple. When in doubt, invoke the aid of the magic number 9. Thus, when dealing with roots, if the divisor will not divide exactly into the other, add 9 to the latter. The root $\frac{7}{8}$ then becomes $\frac{7+9}{8}$ which gives the final root 2.

The same principle applies in subtraction if the final root would otherwise be a negative quantity.

$$\begin{array}{rl} 34 & \text{Root} = 7 \\ 17 & \text{,,} = 8 \end{array}$$

Subtract: 17   ,,   = −1   (add 9 for final root)

                                9
                                —
Final Root   8   8

This method is justified by the principle that the root of the number x is the same as the root of $(x+9)$. No such difficulties arise in addition or multiplication.

The digital process can still be applied to a division, even where the division leaves a remainder. If we divide the number X by the number Y so as to obtain the whole number Z plus a remainder R, then the root of X minus the root of R will equal the root of Y multiplied by the root of Z.

That is:
$$X_r - R_r = Y_r \times Z_r$$

The division of 362 by 179 gives the number 2 and a remainder of 4. Therefore:

$$362_r - 4_r = 179_r \times 2_r$$
or $\qquad 2_r - 4_r = 8_r \times 2_r$
or $\qquad 2 - 4 + 9 = 16_r = 7$

In the third line of working above it was necessary to add 9 to the left-hand side to avoid the appearance of a negative root.

The reason why the digit 9 should apparently have such a remarkable significance as is shown in the last few pages will be found in the study of the various stages of its multiplication. Consecutive multiples of 9 are:

9, 18, 27, 36, 45, 54, 63, 72, 81, 90, etc.

That is, in the tens column the digits 1 to 9 appear in ascending order, whereas in the units column they appear in the reverse order. Thus, for each additional increment of 9, the digit in the tens column increases by 1 and the digit in the unit column decreases by 1. This is because 9 equals $10-1$, so that, in effect, the addition of 9 is the same as adding 10 and deducting 1.

This now gives a further explanation of why the digital root of x is the same as the root of $(x+9)$. When we add 10 to a number, we only add 1 to its digital root, so that if we add 9 to a number, we only add $1-1$, or nought to its root.

It is now instructive to study a table of numbers (A) and

'DIGITAL' ROOTS

their respective roots (B), since a number of interesting points appear:

(A) NUMBERS								(B) ROOTS							
1	2	3	4	5	6	7	8	1	2	3	4	5	6	7	8
2	4	6	8	10	12	14	16	2	4	6	8	1	3	5	7
3	6	9	12	15	18	21	24	3	6	9	3	6	9	3	6
4	8	12	16	20	24	28	32	4	8	3	7	2	6	1	5
5	10	15	20	25	30	35	40	5	1	6	2	7	3	8	4
6	12	18	24	30	36	42	48	6	3	9	6	3	9	6	3
7	14	21	28	35	42	49	56	7	5	3	1	8	6	4	2
8	16	24	32	40	48	56	64	8	7	6	5	4	3	2	1
9	18	27	36	45	54	63	72	9	9	9	9	9	9	9	9
10	20	30	40	50	60	70	80	1	2	3	4	5	6	7	8
11	22	33	44	55	66	77	88	2	4	6	8	1	3	5	7
12	24	36	48	60	72	84	96	3	6	9	3	6	9	3	6
13	26	39	52	65	78	91	104	4	8	3	7	2	6	1	5
14	28	42	56	70	84	98	112	5	1	6	2	7	3	8	4
15	30	45	60	75	90	105	120	6	3	9	6	3	9	6	3
16	32	48	64	80	96	112	128	7	5	3	1	8	6	4	2
17	34	51	68	85	102	119	136	8	7	6	5	4	3	2	1
18	36	54	72	90	108	126	144	9	9	9	9	9	9	9	9

This table shows quite clearly that the digital roots of all multiples of 9 are themselves also 9. They appear in the table as a row of nines. If the table is ruled at these rows, the rulings divide the table into two sections, and it will be seen that in the roots columns, each divided section is identical.

The following facts also appear:

(*a*) In each roots section, the roots forming the diagonals of the section are palindromic. One of these diagonals (from left to right in the first section) represents the square numbers.

(*b*) The roots of all square numbers, in order from 1 upwards, are 1, 4, 9, 7, 7, 9, 4, 1 in that order, and this order is repeated indefinitely. In the same way, although this is not revealed by the table, the roots of all cube numbers are 1, 8, 9, in that order.

(*c*) The root of any multiple of 3 is always 3, 6, or 9.

(*d*) The root of any number not a multiple of 3 is not

similarly restricted. Indeed, within each section of eight numbers, every digit from 1 to 8 appears. But the digits nevertheless conform to a definite pattern within each section and within each row. Thus, multiples of 7 always have their roots in the order 7, 5, 3, 1, 8, 6, 4, 2, 9 repeated indefinitely.

(e) Roots appearing in the first row of each section are repeated in reverse order in the eighth row. The same relationship occurs between the second row and the seventh row, between the third and sixth rows and between the fourth and fifth rows.

Further investigation shows that where numbers can be related to each other as, for example, the numbers in 'shape' series can be related, then their roots also bear a fixed relationship to each other.

The roots of triangular numbers form a regular pattern. The first nine triangular numbers have roots in the following order: 1, 3, 6, 1, 6, 3, 1, 9, 9, and all the higher triangular numbers have roots which repeat this group in exactly the same order.

Number	Root	Number	Root	Number	Root	Number	Root
1	1	55	1	190	1	406	1
3	3	66	3	210	3	435	3
6	6	78	6	231	6	465	6
10	1	91	1	253	1	496	1
15	6	105	6	276	6	528	6
21	3	120	3	300	3	561	3
28	1	136	1	325	1	595	1
36	9	153	9	351	9	630	9
45	9	171	9	378	9	666	9

The roots of hexagonal numbers are even more selective, there being only two possible roots:

Number	Root	Number	Root	Number	Root
1	1	37	1	127	1
7	7	61	7	169	7
19	1	91	1	217	1

To conclude, there follow some further examples of the persistence of 9 as a digital root.

(a) The sum of all the digits from 1 to 9 inclusive is 45 and thus has the root of 9. Similarly the root of the total of any nine consecutive natural numbers is always 9.

(b) Take any number and reverse the digits. If the number thus obtained is deducted from the original number, the root of the difference is always 9. This is self-evident since the root of the original number must be the same as the reversed number and the difference between their individual roots must therefore be 0.

(c) If the separate digits of a number are added together to form a total and this total is deducted from the original number, then the difference will again be a multiple of 9. This follows from the previous example.

(d) Take any number. Reverse its digits to form another number. Square both numbers and subtract the smaller square from the larger. The difference will be a multiple of 9.

Thus:
$$62^2 - 26^2 = 3168 = 9 \times 352$$

This is because the root of 62 is the same as the root of 26 and therefore the root of 62^2 must be the same as the root of 26^2.

5
'Primes and Factors'

A prime number is one which has no factors other than itself and unity. Mathematicians have been searching for centuries to discover a simple test which would enable them, without any great labour, to declare whether or not any specific number is a prime. Many methods have been suggested and proved, but they do not materially shorten the work of testing for primality.

For small numbers the tests of divisibility (see Chapter 6) are of assistance in determining whether numbers have any factors, but for really large numbers the process of elimination is an extremely wearisome business. Nevertheless, although no simple general test is available, the search for such a test has revealed many hitherto unknown number relationships and has therefore not been unproductive.

In testing for factors, there is of course no need to test for any possible factors greater than the square root of the number. For, if x be the nearest whole number to the value of the square root of y and if no numbers smaller than x will divide exactly into y, then no other number greater than x can divide exactly into y so as to give another factor smaller than x.

An ancient pictorial method of segregating prime numbers from other numbers (called Composite numbers) was the use of the Sieve of Eratosthenes. All numbers from 1 to any particular maximum number were written down and subjected to a process of elimination as follows. All even numbers greater than 2 and all numbers ending in 0 or 5 (other than 5 itself) were crossed out. Then, in turn, every third number after 3, every seventh number after 7, every eleventh number after 11, and so on, were crossed out. When the process was complete, the numbers remaining uncrossed were found to be primes. Since this is only a selective method and the number of primes is infinite, the Sieve has only a limited use and gives no clues to a general formula.

The primes less than 100, shown in their respective groups of ten consecutive numbers are:

Between						
1 and 10		1	2	3	5	7	
11 and 20		11		13		17	19
21 and 30				23			29
31 and 40		31				37	
41 and 50		41		43		47	
51 and 60				53			59
61 and 70		61				67	
71 and 80		71		73			79
81 and 90				83			89
91 and 100						97	

It is at once apparent that these numbers do not form any series similar to those we have already met, and are seemingly not related to each other in any fixed way at all.

An ingenious method of finding the factors (if any) of a number was devised by Fermat, who based his calculations on the fact that the factors of $a^2 - b^2$ are $(a+b)$ and $(a-b)$. Thus, if any number can be expressed as the difference between two squares, then its factors may be readily discovered.

The procedure is to calculate the approximate square root of the number to be factorized. The next largest integer is then taken as the starting point. If this integer be called a, and the original number be called N, then if $a^2 - N$ is another perfect square (say b^2), it follows also that $a^2 - b^2 = N$. By this means, the original number N is expressed as the difference between two squares.

It will be seen that a^2 is a certain square number slightly greater than N, but if $a^2 - N$ is not a perfect square then we can substitute $(a+1)^2$ or $(a+2)^2$ or, in fact, any square larger than a^2 as we proceed. Thus, if $a^2 - N$ is not a perfect square, we proceed to substitute $(a+1)$ and progressively higher numbers for a until we reach the point where $(a+x)^2 - N$ is a perfect square (b^2). To simplify this, we may substitute y for $(a+x)$ and show the equation $y^2 - N = b^2$.

From this equation we have:
$$y^2 - N = b^2$$
$$\therefore y^2 - b^2 = N$$
$$\therefore (y-b)(y+b) = N$$
and the factors of N are therefore $(y-b)$ and $(y+b)$.

This may best be illustrated by an example, showing the various stages:

Detail	Numerical example
Number to be factorized:	$N = 323$
Next square greater than N:	$a^2 = 324 \ (=18^2)$
Difference:	$a^2 - N = b^2 = 324 - 323 = 1$
Thus:	$a^2 - N = b^2 = 18^2 - 323 = 1^2$
and:	$a^2 - b^2 = N = 18^2 - 1^2 = 323$
and	$(a+b)(a-b) = N = (18+1)(18-1) = 323$

and this gives 19 and 17 as the factors of 323.

In this example the difference $18^2 - 323$ was a perfect square. If, however, it had not been a perfect square, then we would have formed in turn $(19^2 - 323)$, then $(20^2 - 323)$ and so on until a perfect square did result or was proved to be impossible. It would not, of course, be necessary to calculate the numerical value of each of these identities separately since each square number can be obtained from the previous one by virtue of the relationship shown in Chapter 2. There it was seen that 18^2 was the equivalent of the sum of the first eighteen terms in the series of odd consecutive numbers 1, 3, 5, 7, ... and that the eighteenth term of the series was equal to $(2 \times 18) - 1 = 35$. In order to obtain the value of 19^2 we have only to add the next term in the series (37) to 18^2.

Thus: $18^2 + 37 = 361 = 19^2$

It follows, therefore, that instead of squaring each number

'PRIMES AND FACTORS'

as we employ it as a substitute for a, all we need do is to add progressively higher consecutive odd numbers to the value of a^2-N, until at some stage the total is a perfect square.

The method just outlined is useful where the difference between the factors of a number is relatively small. Where the difference is relatively large, the process is far too protracted.

Another method of factorizing numbers was devised by Euler. This was based on the expression of a number as the sum of two squares in two different ways. For example, the number 221 can be expressed both as (10^2+11^2) and as (5^2+14^2). Not all numbers can be so expressed and, in any case, the method is complicated and brings us no nearer to a general solution.

Although it is not possible to identify every prime immediately, many facts about primes are known. It has, for instance, been proved that the number of primes is infinite. The proof, due to Euclid, is simple and depends upon the basic fact that no two consecutive numbers can have any similar factors. This is so because if x is exactly divisible by y, then $x+1$, when divided by y, will give a remainder of 1.

If instead of x and $x+1$ we substitute $x!$ and $(x!+1)$ where $x! = 1 \times 2 \times 3 \times 4 \times 5 \times \ldots \times x$, the principle remains the same. Therefore $(x!+1)$ cannot have any factors common to $x!$ so that if $(x!+1)$ has any prime divisors at all, then they must be distinct from any number lower than x. Alternatively $(x!+1)$ may itself be prime. One of these propositions must apply so that, in either case, we have proof that there are primes greater than x whatever value we take for x.

In addition to ordinary primes, numbers may be described as relatively prime to each other where they have no common factors. In the same way, the sum of any two relatively prime numbers is also relatively prime to their difference as well as to the numbers themselves. For example, the numbers 16 and 25 are relatively prime. Their sum is 41 and their difference is 9, and the four numbers, 16, 25, 41 and 9 are all relatively prime.

A special interest has been taken over the years in testing for primality those numbers which consist entirely of a repetition of the digit 1. For a long time, the number 11 was the only such number known for certain to be prime, but in 1918 a proof was claimed that the number represented by nineteen digits was also a prime. No other prime of this form has been discovered.

In cases where n (=number of digits) is even, the number is obviously not prime since it will have the factor 11. For cases where n is odd, some factors are known. Where n equals 3, 5 and 7, the factors are respectively:

$$111 = 3 \times 37$$
$$11,111 = 41 \times 271$$
$$1,111,111 = 239 \times 4649$$

On the other hand, the factors, if any, where n equals 23 and 37 are unknown.

Numbers of this nature are of the form $\dfrac{10^n - 1}{9}$, and are, as such, recognizable as being associated with the cycles of recurring decimals (see Chapter 10), and this knowledge enables us to reduce the amount of work in finding their factors. It will be found that the number of digits in a recurring decimal cycle is related to the prime number which, when used as the divisor of unity, generates the cycle. In fact, the number of digits in a cycle is either $(x-1)$ or a factor of $(x-1)$ where x is the prime divisor other than 2 or 5. Now, a number consisting entirely of nines is obviously nine times greater than a number consisting of an equal number of ones, so that any factors of 999 (other than 9 itself) must also be factors of 111. Therefore, in order to find the factors of 111, we can instead find the factors of 999, and the work of finding the latter is simplified by the knowledge that a number of n nines will have a divisor (if any) equivalent to a multiple of n added to 1.

Thus, the factors of 11,111, where $n=5$, are each greater by 1 than multiples of 5. They are 41 and 271. The multiples

'PRIMES AND FACTORS'

of n will always be even multiples where n is odd, since the factors must be odd (being factors of an odd number) and these factors are one greater than the multiples of n. Therefore, when trying to ascertain the factors of the number where $n=5$, we know that the possible divisors can only be:

$$(2 \times 5) + 1 = 11$$
or $$(4 \times 5) + 1 = 21$$
or $$(6 \times 5) + 1 = 31 \quad \text{etc.}$$

and we find the factor 41 after five trial divisions instead of many more trials which would have been necessary if we had had to test for every odd prime lower than 41.

This method can be used in connexion with higher numbers but, although it considerably reduces the work of testing, the amount of work still to be effected is fantastic. Where $n=23$, for example, the possible factors are:

$$(2 \times 23) + 1 = 47$$
$$(4 \times 23) + 1 = 93$$
$$(6 \times 23) + 1 = 139 \quad \text{etc., etc.}$$

If the reader wishes to have any idea of the work involved, he should write down the number consisting of twenty-three ones and divide by the possible factors shown above. He should then consider how many other trial divisions will be necessary before all the other possibilities are exhausted; there is little doubt that the reader would be exhausted first.

An interesting side-issue of the foregoing principle is the fact that a number consisting of n ones is exactly divisible by the number $(n+1)$ if $(n+1)$ is prime, except where $(n+1)$ equals 2, 3 or 5. If the number be represented by X_n where n is the number of digits, then assuming that the number has any factors, these can be written $(M_1 n) + 1$ and $(M_2 n) + 1$, where M_1 and M_2 are the coefficients of some multiples of n.

Thus, $$(M_1 n + 1)(M_2 n + 1) = X_n$$
and, multiplying, $$M_3 n + 1 = X_n$$

But X_n can be written as $10(X_{n-1}) + 1$

so that $M_3n+1=10(X_{n-1})+1$
and $M_3n=10(X_{n-1})$

from which we show that X_{n-1} is a multiple of n, unless n is a factor of 10.

Every prime number greater than 3 differs from a multiple of both 4 and 6 by a difference of 1 (either added or subtracted). Unfortunately it does not follow that *every* number so related to 4 and 6 is necessarily a prime.

The prime

$$19 = (3 \times 6) + 1 = (5 \times 4) - 1$$
whereas
$$55 = (9 \times 6) + 1 = (14 \times 4) - 1$$

and the number 55 is not a prime.

In an endeavour to find a formula for the generating of primes, Euler proposed the form n^2+n+41. This gives a prime number for all values of n up to 39 but obviously fails for $n=41$. The expression n^2-n+41 gives similar results.

It is worthy of note that no triangular number can be a prime since each such number, being of the form $\frac{(n)(n+1)}{2}$, clearly has the factors (n) and $\frac{(n+1)}{2}$.

The usual way of finding the highest common factor of any two numbers is to reduce the numbers to their constituent factors and then to select those factors which are common to both. A more interesting way, discovered by Euclid, is based on the proposition that if two numbers have a common factor then the difference between them also has the same factor. The process is therefore one of expressing one number in terms of the other and then treating the second number and the remainder in a similar way.

In order to find the highest common factor of 6600 and 2424, the number 6600 is first expressed as a multiple of 2424 plus a remainder R_1. The number 2424 is then expressed as a multiple of R_1 plus a new remainder R_2. The number R_1 is then expressed as a multiple of R_2 plus a third remainder R_3.

'PRIMES AND FACTORS'

This process is continued until the first of any two consecutive remainders is an exact multiple of the second.

$$6600 = 2 \times 2424 + 1752$$
$$2424 = 1 \times 1752 + 672$$
$$1752 = 2 \times 672 + 408$$
$$672 = 1 \times 408 + 264$$
$$408 = 1 \times 264 + 144$$
$$264 = 1 \times 144 + 120$$
$$144 = 1 \times 120 + 24$$
$$120 = 5 \times 24 + 0$$

The remainder 120 is an exact multiple of the next remainder 24, and this last number is therefore the highest common factor of the two original numbers.

Provided that a number can be resolved into its constituent basic factors, there is a formula for calculating the number of *different* factors of that number. The number is first reduced to its factors in their lowest forms $(a)^x$ and $(b)^y$ where x and y are the indices representing the powers of each factor included. For example, the number $24 = (2^3)(3^1)$. The formula then employed is based upon the values of the indices of the factors; that is, on x and y. Each index is increased by 1 and they are then multiplied together to give the required total number of factors. For the number 24, where the indices are 3 and 1 respectively, the expression is $(3+1)(1+1) = 8$. These eight factors are 1, 2, 3, 4, 6, 8, 12 and 24.

This formula is of assistance in connexion with the checking of amicable numbers (see Chapter 9), and from it can be derived another formula for calculating the product of all the different factors of any number. If the number be n and the number of different factors be x, then the product of all these different factors is $n^{x/2}$ or $\sqrt{n^x}$. Thus the product of all the different factors of 24 equals 24^4 or 331,776.

The *sum* of all these different factors may also be found rapidly. There are, however, two different methods depending upon the nature of the indices of the basic factors included. Where all the basic factors have the index 1, the sum of all the different factors is found by adding 1 to each of the basic factors and multiplying them together. Where, however, any of the basic factors, a, b, c, etc., have an index greater than 1, as in the number $a^p b^q c^r \ldots$, then the sum of all the different factors is given by

$$\frac{(a^{p+1}-1)\ (b^{q+1}-1)\ (c^{r+1}-1)}{(a-1)\ (b-1)\ (c-1)} \ldots$$

The basic factors of the number 60 are $2^2.3.5$. The sum of all the different factors is therefore

$$\frac{(2^3-1)\ (3^2-1)\ (5^2-1)}{(2-1)\ (3-1)\ (5-1)} = 168$$

This can easily be checked by simple addition, the factors being 1, 2, 3, 4, 5, 6, 10, 12, 15, 20, 30 and 60.

Fermat was responsible for the clarification of many principles affecting number theory, but his expressed opinion that all numbers of the form $2^x + 1$, where $x = 2^n$, are primes, was subsequently proved by Euler to be incorrect. Numbers of this form are called Fermat Numbers for obvious reasons. Where $n = 5$, the expression $2^{2^n} + 1 = 4,294,967,297$ and this number has the factors 641 and 6,700,417.

In 1644, the mathematician Mersenne suggested that in order that numbers of the form $2^P - 1$ should be prime, the only possible values of P, not greater than 257, were 1, 2, 3, 5, 7, 13, 17, 19, 31, 67, 127 and 257, although the number 67 was probably a misprint for 61. In point of fact, not all of these values make $2^P - 1$ prime. Where $P = 257$, the expression is composite.

Two other values of P (89 and 107) have been proved to generate prime numbers. Mersenne's Numbers are closely related to the formula for generating perfect numbers, to which reference is made in a later chapter. For 75 years, the value $P = 127$ gave the largest prime number then known:

$$= 170,141183,460469,231731,687303,715884,105727$$

However, Miller and Wheeler have now found larger ones, the largest being $180(2^{127}-1)^2+1$.

It can be shown that if 2^P-1 is prime, then P must also be prime. If we suppose that P is not prime then it must have at least two factors, a and b, each of which is greater than 1. Then 2^P-1 becomes $2^{ab}-1$, and this expression has the factor 2^a-1, which is neither equivalent to 1 nor to 2^P-1, whence P must be prime. It does not necessarily follow that if P is prime then 2^P-1 is also prime.

Numbers of the form 2^P-2 are divisible by P if the latter is prime. For a long time it was considered probable that, if P were composite, then 2^P-2 could not be divisible by P. This was, however, eventually proved by Sarrus to be incorrect for the value $P=341$.

The subject of primality is not unconnected with the series of consecutive odd numbers. To conclude this chapter, the writer appends a brief survey of the results of individual researches conducted in this connexion, establishing a relationship which is of interest if not of immediate practical use.

SERIES AND PRIMES

In the earlier chapters we have been largely concerned with the use of the series of consecutive odd numbers in the formation of 'shape' or 'power' numbers. This series is also of importance in the construction of every number and has a particular reference to the question of primality.

It was noted that, in order to build up the number n^2, we always required n consecutive numbers and that the same requirement applied for n^3. It may now be stated that any power number of the form n^x can be built up from n consecutive odd numbers irrespective of the value of x. Where n is odd, the central number in the odd series for n^x is equivalent to n^{x-1}. For 5^4 the central number will be $5^{4-1} = 5^3 = 125$, so that:

$$5^4 = 121+123+125+127+129 = 625$$

Where n is even, there is no single central number in the

series; there is instead a pair of numbers and these are respectively 1 less and 1 greater than n^{x-1}. For 4^4, the central pair of numbers are 63 and 65 (that is, $64-1$, and $64+1$) respectively, so that:

$$4^4 = 61+63+65+67 = 256$$

There is a simple reason why this relationship should exist. If an odd number has a factor, then if it be divided by that factor, we can represent it as the sum of its parts, thus:

$$93 = 31+31+31$$

If we deduct 2 from the first of these parts and add 2 to the last, we obtain:

$$93 = 29+31+33$$

which is obviously the same.

In the same way, any even number may be divided into equal parts:

$$64 = 16+16+16+16$$

This time we deduct 3 from the first term and add 3 to the last term; and deduct 1 from the second term and add 1 to the third term, so that:

$$64 = 13+15+17+19$$

This property of divisibility and convertibility to terms of odd numbers is possessed by most composite numbers. An important fact which emerges, however, is that no prime number can be treated in the same way, for the simple reason that they have no factors other than themselves and unity. This suggests a new way for identifying primes.

Let us first examine those numbers which can be built up from the odd number series. We find that all these are composite numbers. For the numbers 1 to 50, we obtain the following results:

Derived from the odd series

1, 4, 8, 9, 12, 15, 16, 20, 21, 24, 25, 27, 28, 32, 33, 35, 36, 39, 40, 44, 45, 48, 49

Not derived

2, 3, 5, 6, 7, 10, 11, 13, 14, 17, 18, 19, 22, 23, 26, 29, 30, 31, 34, 37, 38, 41, 42, 43, 46, 47, 50

It will be seen at once that not all composite numbers can be derived from the odd series, but those which are excluded can be built up from a different series—that of the consecutive even numbers, 2, 4, 6, 8, etc.

We therefore have a method of sorting the primes from the composites; it is in fact a new kind of sieve. If a number cannot be represented either as the sum of a number of consecutive odd or even numbers, then it must be a prime.

This may be more easily appreciated from the following table which shows the sieve in operation:

Numbers which can be built up from:		Numbers which cannot be built up
(a) Odd series	(b) Even series	
1	6	2
4	10	3
8	14	5
9	18	7
12	22	11
15	26	13
16	30	17
20	34	19
21	38	23
24	42	29
25	46	31
27	50	37
28		41
32		43
33		47
35		
36		
39		
40		
44		
45		
48		
49		

All the numbers in the first two columns are composites whereas all the numbers in the third column are primes. This knowledge is, by itself, of little practical use, but it is nevertheless of great importance and leads to a new method whereby specific numbers may be tested for primality. It may indeed reach further, pointing the way to the yet undiscovered general solution.

If we approach the subject from a different angle and assume that a prime might possibly be the sum of a series, we can at once dimiss the possibility where the number of terms in the series is even, since the sum of an even number of terms must be even and is therefore not a prime. It is, in fact, impossible for a prime to be the sum of a series for if it were it would be of the form $\frac{n(a+L)}{2}$ and would have the factors n and $\frac{(a+L)}{2}$.

We are, however, assuming that a prime might be the sum of a series of odd numbers. If we select an odd number, N, then assuming that this can be represented as the sum of a series then:

$$N = \frac{n(a+L)}{2}$$

But $$L = [a+(n-1)d]$$

so $$N = \frac{n}{2}[2a+(n-1)d]$$

and since $$d = 2$$

being the difference between consecutive odd numbers,

then $$N = n(a+n-1)$$
$$= n^2 + (a-1)n$$

whence $n^2 + (a-1)n - N = 0$

We now have the supposed relationship expressed as a quadratic equation. If it has any rational solution at all, this can be found by the general solution formula for quadratic equations, whence:

$$n = \frac{-(a-1) \pm \sqrt{(a-1)^2 + 4N}}{2}$$

$$= \frac{-a+1 \pm \sqrt{a^2-2a+4N+1}}{2}$$

'PRIMES AND FACTORS'

In order that an integral value for n may be derived from the above, it is essential that $a^2-2a+4N+1$ be a perfect square, with a square root of the form $(a+x)$, whence

$$a^2-2a+4N+1=(a+x)^2=a^2+2ax+x^2$$

or
$$2ax+2a=(4N+1)-x^2$$

Sorting out terms and changing signs where necessary, so

$$a=\frac{(4N+1)-x^2}{2(x+1)}$$

Therefore, if there is any integral solution of

$$n^2+(a-1)n-N=0$$

then $2a=\dfrac{(4N+1)-x^2}{(x+1)}$ and from this it will be seen that 2 is a factor of $(4N+1)-x^2$ so that this latter number must be even. Now $(4N+1)$ is odd and therefore x^2 must also be odd. From the same evidence $x+1$ is even and therefore $(4N+1)-x^2$ must be a multiple of 4. Furthermore, in order to satisfy the solution required, the integral number, if any, represented by $\dfrac{(4N+1)-x^2}{(x+1)}$ must be even as it is twice as great as the hypothetical a, which is also required to be integral.

But there can only be a real value for a and n if N can be represented as a sum of a series; that is if it is composite. Therefore if an odd value can be found for x (other than $x=1$) so that $\dfrac{(4N+1)-x^2}{x+1}$ is an even integer, then N is composite. Otherwise, N is prime.

Supposing the number to be composite, it is then an easy matter to evaluate n, which is itself one of the factors. In order to arrive at the above identity, we substituted $(a+x)$ for $\sqrt{(a-1)^2+4N}$, so that reverting to the original equation for the quadratic $n^2+(a-1)n+N=0$, we have:

$$n=\frac{-a+1\pm\sqrt{(a-1)^2+4N}}{2}$$

$$=\frac{-a+1\pm\sqrt{(a+x)^2}}{2}$$

$$=\frac{-a+1\pm(a+x)}{2}$$

$$=\frac{1+x}{2} \quad \text{or} \quad \frac{1-2a-x}{2}$$

The second alternative root is impossible, which leaves the value of

$$n=\frac{1+x}{2}$$

We now take the number 729 as an example. $4N+1$ becomes $4(729)+1=2917$. If 729 is composite, then we should be able to find one or more odd values (other than 1) for x so that

$$\frac{2917-x^2}{x+1}=\text{integral multiple of 2}$$

In fact, we find that:

$$\frac{2917-5^2}{5+1}=482$$

whence $x=5$ and $a=241$. The value of n is then $\frac{1+5}{2}=3$.

The number 729 can therefore be expressed as a series of consecutive odd numbers of three terms, the first term being 241.

$$241+243+245=729$$

It should be noted that other values for x can be found, according to the number of different factors of 729.

For example: $\quad \dfrac{2917-17^2}{(17+1)}=146$

'PRIMES AND FACTORS' 61

whence $a=73$, and $n=9$, so that the number 729 can also be expressed as the sum of nine terms of which the first term is 73:

$$73+75+77+79+81+83+85+87+89=729$$

Yet another entirely new test of primality may be developed from the series of consecutive odd or even numbers. It is much simpler to explain and to employ than is the first method shown above; it is also more practical and therefore of much greater importance. Theoretically it provides a *certain* method of testing *any* number, although the greater the number so also the longer the process. It also reveals what, if any, are the factors.

The method may be stated briefly thus:

(a) If a Composite number is built up from an *even* series, it must itself be even and is thus easily recognizable as composite.

(b) If a Composite number, C, is built up from an *odd* series, it must be of the form of an arithmetical progression
$$C=\tfrac{1}{2}(n)(a+l)$$

But where the common difference is 2
$$l=a+2(n-1)$$

Therefore:
$$\tfrac{1}{2}(n)(a+l)=\tfrac{1}{2}n(a+a+2n-2)$$
$$=n(a+n-1)=C$$

If we substitute values for a

$a=1$	$C=n^2$
$a=3$	$C=n^2+2n$
$a=5$	$C=n^2+4n$
$a=7$	$C=n^2+6n$
$a=9$	$C=n^2+8n$

This may be expressed in the general formula
$$C=n^2+mn$$
where mn is some even multiple of n, and $m=(a-1)$.

Since a is *always odd*, $(a-1)$ or m must always be even. Tests reveal:

$$9 = 3^2$$
$$15 = 3^2 + 2(3)$$
$$21 = 3^2 + 4(3)$$
$$25 = 5^2$$
$$27 = 3^2 + 6(3) \quad = 5^2 + 2(1)$$
$$33 = 3^2 + 8(3) \quad = 5^2 + 8(1)$$
$$35 = 5^2 + 2(5)$$
$$39 = 3^2 + 10(3) = 5^2 + 14(1)$$
$$45 = 3^2 + 12(3)$$

In each case C is either a perfect odd square; or a perfect odd square *plus* an even multiple of that square's root. But what is of much greater significance is the fact that that root (n) is a factor of C, and that $(n+m)$ is another factor.

Thus where
$$C = 27$$
$$n = 3$$
and
$$n + m = 9$$

This relationship is easily proved since
$$C = n(a + n - 1)$$
$$= n(n + m)$$

The procedure for the use of this method is as follows:

Let $C = 11,111$

Nearest odd square less than $C = 11,025$ $(= 105^2)$
$$11,111 = (105^2) + 86$$

The remainder 86 is not a multiple of 105, so that 105 is not a factor of C. A process of elimination, trying successively lower odd squares at each stage, eventually leads to:

$$11,111 = (41)^2 + 9430$$
$$= (41)^2 + 230(41)$$

The factors of 11,111 are therefore 41 and 271 ($= 41 + 230$).

By now it is clear that the series of consecutive odd numbers is of immense importance in the theory of numbers, and it is significant to note that these numbers are themselves directly derived from the series of natural numbers (that is, the series of consecutive integers 1, 2, 3, 4, etc.). Each odd number can in fact be built up from two consecutive natural numbers, whereas even numbers cannot.

Every odd number, other than 1 itself, is clearly one greater than an even number and is therefore of the form $2n+1$. This is the same as $n+(n+1)$ and is therefore the sum of two consecutive natural numbers. No even number can be formed in the same way since, for any pair of natural numbers, one must be odd and the other even so that their sum must also be odd.

6
Divisibility

The behaviour of digits as members of a number, either singly or in groups, is most enlightening when it is necessary to ascertain, without actually dividing out, whether a certain number is exactly divisible by a specified divisor. Tests of divisibility are readily available where the divisor is relatively small, and they are easily explained:

Divisor 1 All numbers are divisible by 1.

Divisor 2 Every number is a multiple of 10, plus its unit digit. Every multiple of 10 is divisible by 2, so that if the last digit of a number is divisible by 2, then so is the number itself.

Divisor 3 A number is divisible by 3 if its digital root is divisible by 3. To prove this, it should be noted that:

$$\left.\begin{array}{r}1000=999+1\\100=99+1\\10=9+1\end{array}\right\} \text{that is, in each case, we have a multiple of 3, plus 1.}$$

Similarly:
$5000 =$ multiple of 3, plus 5
$400 =$,, ,, ,, 4
$20 =$,, ,, ,, 2

So that:
$5420 =$ multiple of 3, $+(2+4+5)$

As, therefore, $2+4+5$ is not divisible by 3, then also 5420 is not.

DIVISIBILITY

Divisor 4   Every number is a multiple of 100 plus its tens and units digits. Every multiple of 100 is divisible by 4, so that if the last two digits of a number when taken together are also divisible by 4, then so is the original number.

Divisor 5   If the last digit of a number is 0 or 5, then it must be divisible by 5, for every number is a multiple of 10, plus its unit digit, and 10 is divisible by 5.

Divisor 6   A number is divisible by 6 if (*a*) it is even, and (*b*) its digital root is divisible by 3. The reasons are of course the same as are shown for the divisors 2 and 3 respectively.

Divisor 7   This is considered with Divisor 13.

Divisor 8   Every number is a multiple of 1000, plus its last three digits. 1000 is a multiple of 8, so that if the last three digits taken together are divisible by 8, then so is the whole number.

Divisor 9   A number is divisible by 9 if its digital root is divisible by 9, for the same reason as shown for Divisor 3.

Divisor 10   By definition, any number ending in nought is made up solely of multiples of 10 without the addition of any units, whence any number having 0 as its last digit must be divisible by 10.

Divisor 11   A number is divisible by 11 if the totals of alternate digits give a multiple of 11 or 0 when one is subtracted from the other. To prove this, it is to be noted that:

$$10,000 - 1 = 9,999$$
$$1,000 + 1 = 990 + 11$$
$$100 - 1 = 99$$
$$10 + 1 = 11$$

so that all the following expressions:

50,000 − 5
7,000 + 7
600 − 6
50 + 5   are multiples of 11.

Therefore 57650 − 5 + 7 − 6 + 5 must result in a multiple of 11, so that 57650 when divided by 11 must give the same remainder as when (5 − 7 + 6 − 5) is divided by 11. In the example given, the number is therefore not divisible.

Divisor 12 The tests of divisibility are the same as for Divisors 3 and 4 in conjunction.

Divisor 7
  ,,   *13* These tests are less well known, but strangely enough are exactly the same for both divisors. The procedure is to split the number to be divided into groups of three digits starting from the right. These groups are then added and subtracted alternately until a number of 3 or fewer digits remains. If this final number is divisible by 7 or 13, then so is the original number; if the final number is 0, then the original number is divisible by both 7 and 13.

Thus 510,300 becomes +510 − 300, reducing to 210. This is divisible by 7 but not by 13, and the original number will be seen to have the same properties. Similarly, the number 947,700 is found to be divisible by 13. Again, the number 151,632 becomes +151 − 632 = −481 and is therefore divisible by 13. It will be seen that the final reduced number may be either positive or negative, but that the result is the same.

These tests became known after the discovery that the geometric series:

$$10^3 \quad 10^6 \quad 10^9 \quad 10^{12} \quad \text{etc.}$$

DIVISIBILITY

where each term is multiplied by 10^3 to obtain the next term, is remarkable for the fact that either the addition to or subtraction from each term of the number 1 will give a multiple of 7 or 13.

$$10^3 + 1 = 1001 \qquad = 7 \times 13 \times 11$$
$$10^6 - 1 = 999999 \qquad = 7 \times 13 \times 10989$$
$$10^9 + 1 = 1000000001 = 7 \times 13 \times 10989011$$

The tests are, of course, useful only in respect of large numbers. A certain type of number may, however, be immediately identified as a multiple of both 7 and 13. Such a number consists of the form '*abcabc*' where obviously $abc - abc = 0$. Thus 176,176 is divisible by both 7 and 13. It should be noted that in the form '*abcabc*', either *a*, *b* or *c* or any two of them may be noughts. 17,017 can be read also as 017,017, and 6006 can be read as 006,006, so that each conforms to the '*abcabc*' form and can readily be identified as multiples of both 7 and 13.

Ordinary division of an apparently complex nature can be very greatly simplified. Many processes in mathematics are made easier by the expedients of borrowing and paying back. In a somewhat different sense, if we wish to divide by a certain number, it may prove simpler to divide by another number and make any necessary adjustments later on or as we proceed. In Chapter 3 it was commented upon that the number 9 has certain peculiarities in multiplication by virtue of its relationship to the number 10. We now have another example in respect of division, for the fact emerges that in order to divide any number by 9, we may instead divide by 10 and make elementary compensating adjustments as we proceed. In fact, these adjustments are so very simple that the whole process can be carried out mentally, by a method first demonstrated by A. H. Russell.

The procedure for dividing a number by 9 is as follows:

(i) Divide the first two digits by 10, giving an answer '*a*' and a remainder '*b*'.

(ii) Write the number '*a*' as the first digit of the quotient in the usual way.

(iii) Add the numbers '*a*' and '*b*' together and carry this total forward to the next digit of the numerator, giving the number '*c*'.

(iv) Divide '*c*' as in (i) above and proceed as before.

The working is:

$$9) 2 \overset{5}{3} \overset{6}{1} \overset{7}{1} 1 (2567 \quad \text{and remainder 8}$$

We first divided 23 by 10. This gave 2 and a remainder of 3. 2 added to 3 gives 5 and this is thus the carry-forward to the next digit, 1. We then divide 51 by 10. This gives 5 and a remainder of 1. 5 added to 1 gives 6 and this is carried forward to the next digit, 1.

It is true, of course, that in the ordinary way, division by 9 can be carried out mentally without recourse to this method, but the value of the foregoing demonstration is in the fact that the system can be applied to much more complicated divisions with equal facility. By its use we can divide by numbers like 99 or 999 in only one or two lines of working.

The usual way of dividing is:

```
99 ) 123456789 ( 12 . . . etc.
     99
     ---
     244
     198
     ---
     465
     etc.
```

and this takes up 13 lines of working.

It is easier to proceed by the other method as already shown, and the following explanation will clarify the reasons.

If we divide by 99 into 100, the answer is 1, plus a remainder of 1, and it will be found that any number of hundreds, when divided by 99, will always give an answer equal

DIVISIBILITY

to the number of hundreds, plus a remainder also equal to the number of hundreds.

$(x)\ 100 \div 99 = x$ and a remainder x
so $\quad 500 \div 99 = 5$,, ,, 5
and $\quad 600 \div 99 = 6$,, ,, 6

From this it follows that:

$[(x)100+y] \div 99 = x$ and a remainder $x+y$
so $\quad (500+34) \div 99 = 5$ and a remainder $5+34\ (=39)$
Now, $\quad 534 \div 100 = 5$ and a remainder of 34
and $\quad 534 \div 99 = 5$ and a remainder of $34+5$

So that to obtain $534 \div 99$, it is possible to divide 534 by 100 and add 5 to the remainder, and for larger numbers, the process is continuous.

The number 123456789, in order to be divided by 99, can now be treated thus. Divide 123 by 100. This gives 1 and a remainder of 23. Write the figure 1 in the answer; add the same number (1) to the remainder (23) and carry 24 forward to the next digit, giving the number 244. Divide by 100. This gives 2 with a remainder of 44. Add 2 to 44 and carry 46 forward and so on.

Fully worked out, this now gives:

```
              24  46  69  3   37  81
    99 ) 1  2  3   4   5  6   7   8  9 (
            ─────────────────────────────
             1  2   4   6  0   3   8  (remainder
                            1              19+8=27)
            ─────────────────────────────
             1  2   4   7  0   3   8
```

It will be seen that when we reach the sixth digit in the original number we have, with the 69 brought forward, the number 696. This divided by 600 gives 96 remainder. This added to the digit (6) in the answer gives 102. As this exceeds 99 the procedure here is to deduct 99 from the remainder, adding 1 to the answer digit and carrying forward a reduced remainder of $102-99=3$.

The same principle can be applied to many numbers in a

similar manner. Thus, to divide by any number ending in 9, divide by that number increased by 1, and proceed as before. In other words, to divide by 19, 39 or 79, divide instead by 20, 40 or 80 respectively and make the necessary adjustments at each stage.

If the divisor ends with the digit 8, the principle may still be employed, but with a difference.

$$100 \div 98 = 1 \quad \text{and} \quad 2 \times 1 \text{ over}$$
$$600 \div 98 = 6 \quad \text{and} \quad 2 \times 6 \text{ over}$$
$$x(100) \div 98 = x \quad \text{and} \quad 2x \quad \text{over}$$

Therefore, to divide by 98, we can still divide by 100, but this time each quotient must be *doubled* before it is added to the remainder to form the carry-forward.

To divide 123456789 by 98, proceed as follows. Divide 123 by 100, giving 1 and 23 over. Write the quotient 1 in the answer. Double this quotient and add to 23 (=25) and carry this amount forward. Divide 254 by 100 and proceed as before.

```
              25  58  95  74  61  30
    98 ) 1 2  3   4   5   6   7   8   9 (
            ─────────────────────────
             1   2   5   9   7   6   3  (and 15 over)
```

Similarly, to divide by 97, we can divide by 100 and obtain carry-forward figures by *trebling* each quotient before adding in the remainder. Again, to divide by 96, divide instead by 100 and *quadruple* each quotient before adding the remainder to find the carry-forward.

If the divisor ends with the digit 5 or a smaller digit, the same principle can be utilized if both divisor and dividend are both multiplied by 2 or 4. Thus to divide 1234 by 5, divide instead 2468 by 10. Similarly, to divide 1234 by 22, multiply both by 4 and so divide 4936 by 88 (i.e. divide by 90 and make the necessary adjustments.)

To divide by 101 it should be noted that the divisor can be written $100+1$, in the same way as 99 can be written $100-1$. It is logical therefore that, to divide by 101, it should be possible to divide by 100 and make adjustments.

Thus,
$$400 \div 101 = 3$$
and remainder of $97 \; (= 100 - 3)$
$$x(100) \div 101 = (x-1)$$
and remainder of $[100-(x-1)]$ or $[101-x]$

So, to obtain the desired result, the process is to divide by 100 and then *deduct* each quotient from the remainder before carrying the latter forward.

As division is really another aspect of multiplication (since to divide a number by x is the same as multiplying that number by $\frac{1}{x}$) it is not surprising that similar methods exist for curtailing the work of multiplying. For example, to multiply any number by 99 it is much simpler to multiply by 100 and then to deduct the original number to give the required answer

7

Multiplication with a Difference

In addition to the ordinary methods of multiplying or dividing, there are a number of other ways of performing the same operations and these at first glance appear to have no obvious connexion with accepted methods at all. Some of the primitive methods appear to be complex in their simplicity, but the inherent paradox is easily reduced to a pattern when related to mathematical fundamentals.

An ancient method of multiplying together two numbers *larger* than 5 enables this to be done by simple operations with numbers *smaller* than 5. Thus, if we select the two numbers 8 and 9, their product can be found by:

(a) Deduct 5 from each, leaving 3 and 4; add these two numbers, giving 7.

(b) Deduct from 5 the respective residues in (a), that is 3 and 4, leaving 2 and 1; and multiply these two numbers together, giving 2.

(c) Then the answer, consisting of two digits, will have in its tens column the number in (a), and in its units column the number in (b), giving 72.

The justification for this method can be seen by substituting x and y for the two numbers to be multiplied. The stages are then taken to give the formula:

$$(x-5)10 + (y-5)10 + (5-x+5)(5-y+5)$$
$$= 10x - 50 + 10y - 50 + 100 - 10x - 10y + xy$$
$$= xy.$$

Another early method was to proceed as follows. Form a quadrilateral divided into small squares, the number of the

MULTIPLICATION WITH A DIFFERENCE

squares (and therefore the size of the quadrilateral) depending upon the size of the numbers to be multiplied. Each square is then bisected diagonally. The next step is to write one of the numbers along the quadrilateral, allowing one digit to each square, and in the same way to write the other number along the right-hand side.

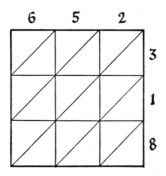

Now in each small square we write down the product of the digits along each side of the quadrilateral. Thus in the first square we write the number 18 (i.e. 3 ×6), in the second

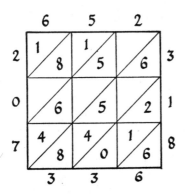

square 15 (i.e. 3 ×5) and so on until all the squares are filled. When a product consists of one digit, it is written to the right of the diagonal bisecting the square. When it consists of two digits, the first digit is written to the left, and the second digit is written to the right of the diagonal. Next, the digits within the squares are added diagonally irrespective of the squares, and the results are written in order backwards along the bottom and continuing upwards along the left-hand side of the quadrilateral. As in ordinary addition, tens are carried forward to the next diagonal.

The answer, 207336, can then be read off starting at the top of the left-hand side and continuing along the bottom of the quadrilateral.

There is another old method, whereby any two numbers may be multiplied together, which involves only actual multiplication by the number 2. The procedure, known as Russian Multiplication, is as follows. Form two columns, headed respectively by the two numbers to be multiplied; it does not matter which number goes in which column. The number in the left column is progressively divided by 2, any remainders being ignored. The number in the right column is in equal stages progressively multiplied by 2. When the number 1 is reached in the left column, both columns are ruled off. All even numbers in the left column are crossed out, as also are the numbers in the same lines in the right column. The remaining numbers in the right column are then added and their total gives the answer to the multiplication.

397 × 196 would be calculated:

397	196
~~198~~	~~392~~
99	784
49	1568
~~24~~	~~3136~~
~~12~~	~~6272~~
~~6~~	~~12544~~
3	25088
1	50176
Total	77812 = 397 × 196

MULTIPLICATION WITH A DIFFERENCE

Superficially, the remarkable things about this method of calculation are (*a*) in dividing, remainders are ignored, and (*b*) the even numbers in the left column are always deleted; yet the answer is always correct. This is to some extent explained by the fact that all even numbers are composed of varying powers of 2, and that all odd numbers are, in reality, even numbers plus 1. This will more clearly be seen from the following table.

$$
\begin{aligned}
1 &= 1 & 5 &= 2^2 + 1 \\
2 &= 2 & 6 &= 2^2 + 2 \\
3 &= 2 + 1 & 7 &= 2^2 + 2 + 1 \\
4 &= 2^2 & 8 &= 2^3
\end{aligned}
$$

As the numbers increase, new factors appear at certain stages. These factors are in order 1 (or 2^0), 2 (or 2^1), 2^2, 2^3, etc. A few examples of simple multiplication are now helpful. In

(*a*) 2×10, the columns will be

```
2    10
1    20
```

To obtain the answer the *second* term only is taken.

(*b*) for 3×10, we have:

```
3    10
1    20
```

To obtain the answer,
the *first and second* terms are taken.

(*c*) for 14×10, we have:

```
14    10
 7    20
 3    40
 1    80
```

To obtain the answer,
the *second, third and fourth* terms are taken.

(d) for 24 × 10, we have:

$$\begin{array}{cc} 24 & 10 \\ 12 & 20 \\ 6 & 40 \\ 3 & 80 \\ 1 & 160 \end{array}$$

To obtain the answer, the *fourth and fifth* terms are taken.

In each case the terms taken in the right column coincide with the odd numbers in the left column.

When multiplying by 24 (as in the last example) we are really multiplying by (2^4+2^3) and this reveals a relationship between (*a*) the powers of 2 of which the multiplier is composed, and (*b*) the numbers of the terms to be taken in the other column. Thus if the multiplier is 2^n, we take the $(n+1)$th term in the right column. If the multiplier is (2^n+2^x) then we take the $(n+1)$th and the $(x+1)$th terms in the right column. In other words, the numbers of the terms to be taken are always 1 greater than the indices representing the powers of 2 included in the multiplier.

Thus, to multiply by 72 this must first be reduced to its components of powers of 2. This is done by taking the highest power of 2 included (i.e. $2^6=64$) and expressing the remainder in the same way. Thus $72=2^6+2^3$. The indices of the factors are 6 and 3, so that the terms to be taken in the right column will be the seventh $(6+1)$ and fourth $(3+1)$, so:

$$\begin{array}{ccl} 72 & 60 & \text{(1st term)} \\ 36 & 120 & \text{2nd ,,} \\ 18 & 240 & \text{3rd ,,} \\ 9 & 480 & \text{4th ,,} \\ 4 & 960 & \text{5th ,,} \\ 2 & 1920 & \text{6th ,,} \\ 1 & 3840 & \text{7th ,,} \end{array}$$

The fourth and seventh terms together total 4320 and this equals 72 × 60.

When we multiply by an odd number, say 17, we are in fact multiplying by 2^4+1. Hence the first and fifth terms

MULTIPLICATION WITH A DIFFERENCE

are taken. The first term in the right column will always be taken when the multiplier in the left column is odd. But the first term in the left column is always the multiplier itself, so that when the multiplier is odd, the first term in the left column is odd. So much is clear. What is not so clear is why the subsequent terms which must be taken in the right column are always on the same lines as odd numbers in the left column. The last term is always included for the last term in the left column is always 1 and therefore odd. The remaining problem is how the intervening included terms 'know' they have to be odd.

The general theory may be explained thus: Take any figure (say 99) and split it into powers of 2 ($99=64+32+2+1=2^6+2^5+2+1$). Now instead of writing 99 in the left column, write the factors there and watch what happens: when we divide successively by 2.

Left Column	Right Column	Term
$99 = \begin{cases} 2^6 \\ 2^5 \\ 2 \\ 1 \end{cases}$	2	1st
$49 = \begin{cases} 2^5 \\ 2^4 \\ 1 \end{cases}$	4	2nd
$24 = \begin{cases} 2^4 \\ 2^3 \end{cases}$	8	3rd
$12 = \begin{cases} 2^3 \\ 2^2 \end{cases}$	16	4th
$6 = \begin{cases} 2^2 \\ 2 \end{cases}$	32	5th
$3 = \begin{cases} 2 \\ 1 \end{cases}$	64	6th
$1 = 1$	128	7th

In the left column the plus-signs are omitted between the

factors for convenience: they should be understood to be there. If 99 is represented by 2^6+2^5+2+1 and is divided by 2 the result will be 2^5+2^4+1, ignoring the remainder (as in the second line). Eventually, as we divide by 2, each power of 2 will be reduced to 1, and it is the lines in which each power is so reduced that contain, in the right columns, the terms which must be added. Thus each factor is reduced to 1 as follows:

Factor	Term in which reduced
2^6	7th
2^5	6th
2	2nd
1	1st

It is clear that the 7th, 6th, 2nd and 1st terms will be odd in the left column since each of these is a multiple of 2, plus 1. And it is also clear that the relative terms in the right column are those which have to be added. Omitting irrelevant terms we have:

99	2
49	4
3	64
1	128
	Add: 198 $(=2\times 99)$

The reason for this is that as each left-hand factor is reduced to 1, the right-hand term is the result of multiplying the original number by the factor reduced. Thus the factor 2 is reduced to 1 in the 2nd term and the right-hand figure has been multiplied by 2. Similarly, 2^6 is reduced to 1 in the seventh term and the right-hand figure has been multiplied by 2^6. So, if we add the 1st, 2nd, 6th and 7th terms in the right column we are in fact adding:

$$(1\times 2)+(2\times 2)+(2^5\times 2)+(2^6\times 2)$$

or $\qquad 2(1+2+2^5+2^6)=2\times 99$

MULTIPLICATION WITH A DIFFERENCE

The principle of multiplication and division in successive stages by the number 2 can be applied to division in exactly the same way since the division of one number by another is the same as multiplying the first by the reciprocal of the second number. Thus $99 \div 7$ can also be represented as $99 \times \frac{1}{7}$ and we can proceed as before:

$$
\begin{array}{cc}
99 & \frac{1}{7} \\
49 & \frac{2}{7} \\
\cancel{24} & \cancel{\frac{4}{7}} \\
\cancel{12} & \cancel{1\frac{1}{7}} \\
\cancel{6} & \cancel{2\frac{2}{7}} \\
3 & 4\frac{4}{7} \\
1 & 9\frac{1}{7} \\
\hline
99 \div 7 = & 14\frac{1}{7} \\
\hline
\end{array}
$$

This method is, of course, quite impracticable for present-day needs but was of very real assistance at a time when men found great difficulty in multiplying large numbers. It is included here mainly because it makes use of the knowledge of how numbers are made up and serves to emphasize the nature of that structure.

It is a well-known fact that the squaring of a number consisting of a whole number and the fraction $\frac{1}{2}$, can be carried out by adding $\frac{1}{4}$ to the product of the whole number and the next whole number.

Thus,
$$(7\tfrac{1}{2})^2 = 7 \times (7+1) + \tfrac{1}{4} = 56\tfrac{1}{4}$$

This is because:
$$(x+\tfrac{1}{2})^2 = x^2 + x + \tfrac{1}{4}$$
$$= x(x+1) + \tfrac{1}{4}$$

So that
$$(7+\tfrac{1}{2})^2 = 7(7+1) + \tfrac{1}{4} \text{ as above}$$

The same principle can be applied to all numbers ending in 5.

Thus,
$$35 = 10 \times 3\tfrac{1}{2}$$
Therefore
$$35^2 = 10^2 \times (3\tfrac{1}{2})^2$$
$$= 100 \times [(3 \times 4) + \tfrac{1}{4})]$$
$$= 100 \times 12\tfrac{1}{4} = 1225$$

This result can be seen at a glance. Multiply 3 by 4 and call the answer hundreds; for the last two digits, square the number 5.

Thus,
$$35^2 = 100(3 \times 4) + 5^2$$
$$= 1200 \quad + 25 = 1225$$

Any two numbers, each consisting of a whole number and a fraction, can be multiplied in the same way provided (*a*) the whole number is the same in each case, and (*b*) the two fractions together add up to 1.

For example
$$3\tfrac{1}{4} \times 3\tfrac{3}{4} = (3 \times 4) + (\tfrac{1}{4} \times \tfrac{3}{4}) = 12\tfrac{3}{16}$$

This procedure can be extended to numbers not having fractions, where the 'tens' digits are the same, and the 'ones' digits together add up to 10.

Thus,
$$31 \times 39 = 100(3 \times 4) + (1 \times 9)$$
$$= 1209$$

8
Logarithms and Trigonometrical Ratios

The system of logarithms is based upon the fundamental fact that every integer can be expressed as a power of any other integer, the latter then being the base to which the logarithms are calculated. If any two numbers to be multiplied together are first converted to powers of the same base number, then the special rules affecting their exponents apply, and the task of multiplying is converted to one of addition. Ordinary logarithm tables in everyday use are calculated to the base of 10.

The numbers 10, 100, 1000 and 10,000 form a geometric progression as follows:

Progression:	10^1	10^2	10^3	10^4
Exponent of term:	1	2	3	4
Progression:	10	100	1000	10000

The two progressions are of course the same. The terms in the third row are converted, in the first row, to powers of 10 and the exponents of these power numbers are shown in the middle row. In order to multiply 10 by 1000 (that is, the first term by the third) we find the answer in the fourth (that is, $1+3$) column.

This is the skeleton framework of our logarithm tables. The terms in the second row are the logarithms of the terms in the bottom row and, conversely, those in the bottom row are the antilogarithms of those in the second row. The complete tables as now used were compiled by filling in the gaps between 10 and 100 and higher powers of 10 by converting the intervening integers to approximate powers of 10. For example, 100 equals 10^2 and the logarithm of 100 is therefore 2. In the same way, the number 3 is approximately equal to $10^{.4771}$ and the number 79 is approximately equal to $10^{1.8976}$, so that the logarithms of 3 and 79 are ·4771 and 1·8976 respectively. Emphasis is laid upon the fact that although these equivalents are only approximate and not exact, the margin of error is so small as to be negligible, except where results to many more significant figures are required.

There are two special points worthy of note. Firstly, the logarithm of the base number itself is always 1, because $a^1 = a$, and it follows that the logarithm of a is 1. Secondly, the logarithm of 1 is always 0, whatever base is used, for $a^0 = 1$, irrespective of the value of a.

At first sight the second example may seem absurd, since it means that $1^0 = 2^0 = 3^0$, etc., but examination of the meaning of a^0 removes the apparent absurdity. For a^0 is the same as a^{n-n} and this is the same as a^n divided by a^n, and this is clearly equal to 1 for any value of a. This will be more clearly understood if we take a^n to be $1 \times a^n$ (which is obviously the same) and, instead of saying that a^n represents the product of a taken n times, we say that it represents the number 1 multiplied by a taken n times. This is a very different description. It follows then that a^0, being the same as $1 \times a^0$, means 'take the number 1 and do not multiply it by any number of a's at all'. In other words, the number 1 is not to be multiplied at all and it must therefore remain 1.

The use of logarithms for multiplication and division reduces the working of the process considerably, especially where there are a number of simultaneous operations. Their use is, however, even greater in the extraction of roots, for not only can this be done in a matter of seconds but, what is more important, any root of any degree may be found with equal ease. There is, for instance, no practical way of finding the fifth root of a number by ordinary arithmetical pro-

LOGARITHMS AND TRIGONOMETRICAL RATIOS

cesses, yet it can be calculated merely by dividing the number's logarithm by five and then treating the result as the antilogarithm of the root, which may easily be found from the tables.

The invention of logarithms by Napier was undoubtedly influenced, if not directly encouraged, by the discovery of certain relationships between the numerical values of trigonometrical ratios and is, itself, an example of the ways in which the purely 'number' branch of mathematics is related to the constructional branches such as geometry and trigonometry. We have already mentioned Pythagorean numbers in Chapter 2; these were found to be related to the geometric construction of right-angled triangles, and it is from the trigonometrical ratios of these same triangles that was first discovered a way of dealing with mathematical processes akin to that upon which the theory of logarithms is based.

A ratio is the result of comparing two quantities of the same kind by dividing one by another and expressing the result as a fraction. In trigonometrical ratios, the two quantities compared are the lengths of the sides of right-angled triangles and the results obtained are expressed as relatives of the angles contained between the two sides compared in any one ratio. These ratios are called sines, cosines and tangents, and for any triangle ABC, such that the angle C is a right-angle, the respective ratios are calculated as follows:

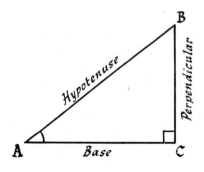

$$\text{Sine } A = \frac{\text{Perpendicular}}{\text{Hypotenuse}}$$

$$\text{Cosine } A = \frac{\text{Base}}{\text{Hypotenuse}}$$

$$\text{Tangent } A = \frac{\text{Perpendicular}}{\text{Base}}$$

If the angle C remains a right-angle, it follows that, as the angles of a triangle always total 180°, the angles B and A must always add up to 90°. Thus, as A is increased by 1°, so the angle B is decreased by 1°, and in short the angles A and B are always complementary. It follows from this that the cosine of angle A is the same as the sine of angle B, being equivalent to $\dfrac{\text{Side } AC}{\text{Side } AB}$, for the side AC is the base side in relation to the angle A, but is the perpendicular side in relation to the angle B.

Since there is such a close relationship between sines and cosines, it is not surprising to find that they can be related in many different ways. In particular, there are two expressions with which we are concerned here (proofs being available in any book on trigonometry). The two expressions are:

(i) $\text{Sin}(A+B) = \text{Sin } A \cdot \text{Cos } B + \text{Sin } B \cdot \text{Cos } A$.

(ii) $\text{Sin}(A-B) = \text{Sin } A \cdot \text{Cos } B - \text{Sin } B \cdot \text{Cos } A$.

If these two expressions are added together, we obtain the result:

(iii) $\text{Sin}(A+B) + \text{Sin}(A-B) = 2 \cdot A \cdot \text{Cos } B$

so that

(iv) $\text{Sin } A \cdot \text{Cos } B = \frac{1}{2}[\text{Sin}(A+B) + \text{Sin}(A-B)]$

Therefore, if it is required to multiply together two numbers which are the equivalents of $\text{Sin } A$ and $\text{Cos } B$ respectively, this can be achieved by calculating the value of the expression $\frac{1}{2}[\text{Sin}(A+B) + \text{Sin}(A-B)]$. For example, in order to multiply ·2588 by ·9848, we proceed as follows:

$$·2588 = \text{Sin } 15°$$
$$·9848 = \text{Cos } 10°$$

LOGARITHMS AND TRIGONOMETRICAL RATIOS

so that $·2588 \times ·9848 = \text{Sin } 15° \times \text{Cos } 10°$, and this gives the following values to be substituted in (iv) above:

$$A = 15°$$
$$B = 10°$$
$$A + B = 25°$$
$$A - B = 5°$$

and we obtain:

$$\text{Sin } 15° . \text{Cos } 10° = \tfrac{1}{2}(\text{Sin } 25° + \text{Sin } 5°)$$

From the tables of sines, we find:

$$\text{Sin } 25° = ·4226$$
$$\text{Sin } 5° = ·0872$$

so that

$$\tfrac{1}{2}(\text{Sin } 25° + \text{Sin } 5°) = \tfrac{1}{2}(·4226 + ·0872) = ·2549$$

therefore

$$·2588 \times ·9848 = ·2549$$

To multiply these same two numbers by the use of logarithms, we have:

$$\text{Log } ·2588 = \bar{1}·4130$$
$$\text{Log } ·9848 = \bar{1}·9934$$

$$\text{Add: } \quad \bar{1}·4064$$

$$\text{Antilog } \bar{1}·4064 = ·2549$$

Logarithms calculated to one base number may very easily be converted to another base. The logarithm of the number 4 to the base of 2 is obviously 2, since $4 = 2^2$. On the other hand, the logarithm of the number 4 to the base of 10 is ·6021.

Thus:
$$\text{Log}_2 4 = 2·0$$
$$\text{Log}_{10} 4 = ·6021$$

If we divide the first logarithm (to the base of 2) by the second logarithm (to the base of 10), we obtain the ratio 3·322, and this ratio proves to be the logarithm of the second base number when calculated to the first base. That is:

$$\text{Log}_2 10 = 3·322$$

Therefore, in order to obtain the logarithm to base 10 of any number, we divide the logarithm to base 2, by the constant 3·322. This can be proved as follows. Let N be any number, so that $N=10^n$.

Then: $\qquad\qquad\qquad n = \text{Log}_{10} N$ . . . . (i)

But $\qquad\qquad\qquad N = 10^n \quad \text{and} \quad 10 = 2^{3·322}$

Therefore $\qquad\qquad\quad N = 2^{3·322n}$

and $\qquad\qquad 3·322n = \text{Log}_2 N$

But $\qquad\qquad\qquad n = \text{Log}_{10} N \quad$ (from (i) above)

so $\qquad\quad 3·322 \, \text{Log}_{10} N = \text{Log}_2 N$

and $\qquad\qquad \text{Log}_{10} N = \dfrac{\text{Log}_2 N}{3·322}$

In general the relationship becomes:

$$\text{Log}_a N = \dfrac{\text{Log}_b N}{\text{Log}_b a}$$

where b is the original base number and a is the base to which the original logarithm is to be converted.

9
Perfect Numbers and Some Oddities

There are certain numbers which individually or as members of a group possess properties peculiar to themselves. Some of these are related to each other, but others have no apparent relation to any other numbers with similar properties.

The easiest to classify are the perfect numbers. A perfect number is one which is itself equal to the sum of all its factors (including 1 as a factor). The lowest of such numbers is 6, whose factors (1, 2 and 3) also total 6. Only twelve of these numbers are known. The first four are 6, 28, 496 and 8128; the fifth is 33,550,336 and the others are much greater. All of these known numbers end either with the digit 6 or the two digits 28; and so far no odd perfect numbers have been discovered, unless we accept 1 as such.

Euclid proved that any number of the form $(2^n)(2^{n+1}-1)$ is a perfect number when the factor $(2^{n+1}-1)$ is a prime number. Thus 496 can be factorized to $(2^4)(2^5-1)$ or $(16) \times (31)$, and since 31 is a prime, then 496 is perfect. It will be observed that the known perfect numbers become much less frequent as numbers grow larger. This is because the primes on which they are based become more rare.

There is a way of building up perfect numbers without having direct regard to Euclid's formula. The factors of 6 are 1, 2 and 3; and these are in the form of an arithmetical progression. This provides a clue that the other perfect numbers may also prove to be equivalent to the sums of arithmetical progressions. This is, in fact, the case; and each of these progressions is inter-related to the others.

The first five perfect numbers can now be shown:

$$6 = \text{Sum of progression 1 to 3 inclusive}$$
$$28 = \text{ ,, ,, ,, 1 to 7 ,,}$$
$$496 = \text{ ,, ,, ,, 1 to 31 ,,}$$
$$8128 = \text{ ,, ,, ,, 1 to 127 ,,}$$
$$33{,}550{,}336 = \text{ ,, ,, ,, 1 to 8191 ,,}$$

The last term is, in every case, a prime (being of the same form as in Euclid's formula), and the only other factors of the perfect numbers are 1 and varying powers of the number 2.

The relationship between each progression is as follows. Once the last term (l_1) of an earlier progression is known, the last term (l_2) of the next progression is found by multiplying l_1 by 2 and adding 1. So, $l_2 = 2(l_1) + 1$; but only provided the resulting l_2 is prime. For example, as between the two numbers 6 and 28, the last terms of their respective progressions are 3 and 7 (that is, $l_1 = 3$; $l_2 = 7$; and therefore $l_2 = 2(l_1) + 1$). If, however, the resulting l_2 is not a prime then the procedure of doubling and adding 1 (at each stage) is continued until a prime does arise. This prime, when reached, will then be the last term of the next perfect number's progression.

This explains the gap between the fourth and fifth perfect numbers. The last term of the progression for the fourth number is 127. The successive doubling up and addition of unity at each stage give the following:

(*Fourth number*) Last term of A.P. = 127	Prime
Double and add 1 = 255	Not Prime
,, ,, ,, 1 = 511	,, ,,
,, ,, ,, 1 = 1023	,, ,,
,, ,, ,, 1 = 2047	,, ,,
,, ,, ,, 1 = 4095	,, ,,
,, ,, ,, 1 = 8191	Prime

The last term of each progression also bears a direct relationship to its perfect number other than as a mere factor, and it is instructive to compare the structure of a perfect number (regarded as the sum of an arithmetical progression) with its factorial composition.

PERFECT NUMBERS AND SOME ODDITIES 89

The perfect number 496 is the sum of the series 1 to 31, and its various terms may be re-grouped as follows:

1 2 3 4 5 6 7 8 9 10 11 12 13 14 15 16
31 30 29 28 27 26 25 24 23 22 21 20 19 18 17

Each of the first fifteen groups totals 32, leaving the number 16 unpaired. The total of all the terms is therefore $(15 \times 32) + 16$ or $(15\frac{1}{2} \times 32)$.

On the other hand, the factors of 496 are 1, 2, 4, 8, 16, 31, 62, 124, 248, and these can be re-grouped thus:

$$\begin{aligned}
1+31 &= 32 = 1 \times 32 \\
2+62 &= 64 = 2 \times 32 \\
4+124 &= 128 = 4 \times 32 \\
8+248 &= 256 = 8 \times 32 \\
16 &= 16 = \tfrac{1}{2} \times 32 \\
\hline
\text{Total } & 15\tfrac{1}{2} \times 32
\end{aligned}$$

In each case the number 496 is shown as being $(15\frac{1}{2} \times 32)$ or $\dfrac{31 \times 32}{2}$. In other words, the last term of the progression for a perfect number x must, when added to 1, give a number n so that

$$\frac{n^2 - n}{2} = x$$

Now if
$$x = \frac{n^2 - n}{2}$$

then
$$2x = n^2 - n = n(n-1)$$

As x grows larger, so $(n-1)$ becomes more nearly n, so that $2x = n^2$ approximately. Therefore the value of n may be found as being very nearly equal to the square root of $2x$.

For example, if $x = 496$, the square root of twice this number is between 31 and 32, so that $n = 32$, and $(n-1) = 31$.

The fact that these known perfect numbers are equal to the sums of arithmetical progressions of the forms 1, 2, 3, 4, etc., means that they must also be triangular numbers. They are in fact particular triangular numbers, and it does not follow that all triangular numbers are perfect.

Somewhat similar to perfect numbers are those known as amicable numbers. These are pairs of numbers, each of which is equal to the sum of all the factors of the other. Thus, the sum of all the factors of the number 220 is equal to 284, while the sum of all the factors of 284 is equal to 220. The next pair of amicable numbers are 1184 and 1210, and other examples are 17296 and 18416. The other known amicable numbers are much greater.

In another category are the Automorphic numbers, the special property of which is that their digits reappear as the last digits of their squares and higher powers. Examples are 25 and 76.

Palindromic numbers are those which are the same either read forwards or backwards. The most simple form of palindromic numbers is the one containing a number of identical digits. Some palindromic numbers are themselves derived from others, as is shown by the following table of the squares of numbers consisting entirely of the repetition of one digit.

$$11^2 = 121$$
$$111^2 = 12321$$
$$1111^2 = 1234321$$
$$11111^2 = 123454321$$
$$111111^2 = 12345654321$$
$$1111111^2 = 1234567654321$$
$$11111111^2 = 123456787654321$$
$$111111111^2 = 12345678987654321$$

Higher squares, however, give numbers of a different pattern. Another interesting series is that in which the palindromes form the centres of the numbers shown above:

$$232 = 4 \times 58$$
$$23432 = 4 \times 5858$$
$$2345432 = 4 \times 586358$$
$$234565432 = 4 \times 58641358$$
$$23456765432 = 4 \times 5864191358$$
$$2345678765432 = 4 \times 586419691358$$
$$234567898765432 = 4 \times 58641974691358$$

Each of these numbers is a multiple of 4 and possesses another factor which can be derived from the factors of the previous numbers. To obtain the number 5858, we insert 85 in the centre of the previous number 58; to obtain 586358, we insert 63 in the centre of 5858. Thus, at each stage, two digits are added to the centre of the preceding palindrome. These additional digits are themselves in series, being:

$$85 \quad 63 \quad 41 \quad 19 \quad 96 \quad 74$$

Each of these is 22 less than the preceding term. (Note: add 99 to 19 to start afresh. This gives 118, and 22 deducted from this gives the next term 96.)

Some palindromes are perfect squares, such as the number 698896, the square of 836; and, finally, it is interesting to note that 12345679×99999999 gives the palindrome 12345678-87654321.

There now follow some further examples of number relationships which cannot be conveniently classified.

1. Some pairs of numbers are related thus: Two numbers are multiplied together (giving answer A). The digits of the two numbers are then reversed and the two new numbers are multiplied together (giving answer B).

$$312 \times 221 = 68952 \text{ (answer } A\text{)}$$
$$213 \times 122 = 25986 \text{ (answer } B\text{)}$$

The two answers are then found to be reversals of each other.

2. Certain numbers can be represented as the aggregate of the cubes of their individual digits.

$$153 = 1^3 + 5^3 + 3^3$$
$$370 = 3^3 + 7^3 + 0^3$$
$$371 = 3^3 + 7^3 + 1^3$$
$$407 = 4^3 + 0^3 + 7^3$$

3. Some square numbers have as their square roots the aggregate of their own digits, taken in pairs:

$$3025 = 55^2 \quad \text{and} \quad 30 + 25 = 55$$
$$9801 = 99^2 \quad \text{and} \quad 98 + 01 = 99$$
$$2025 = 45^2 \quad \text{and} \quad 20 + 25 = 45$$

4. Certain other square numbers have roots which, when reversed and squared, give the reverse of the original numbers.

$$12^2 = 144 \text{ and } 21^2 = 441$$
$$13^2 = 169 \text{ and } 31^2 = 961$$

5. Some numbers when multiplied give the same digits but in reverse order.

$$2178 \times 4 = 8712$$
$$1089 \times 9 = 9801$$
$$4356 \times 1\tfrac{1}{2} = 6534$$

It will be noted that all these numbers are multiples of 1089.

6. Some numbers, taken together and added, give the same result as when they are multiplied.

$$2 \times 2 = 4 \text{ and } 2 + 2 = 4$$
$$1\tfrac{1}{2} \times 3 = 4\tfrac{1}{2} \text{ and } 1\tfrac{1}{2} + 3 = 4\tfrac{1}{2}$$

Apart from the first example given above, one of the numbers in each pair must always be in the form of a fraction, and there is a simple rule which enables such pairs to be generated.

If x be one of the numbers, then the other will be a fraction of the form $\dfrac{x}{x-1}$.

Adding these two we get:

$$x + \frac{x}{x-1} = \frac{x^2 - x + x}{x-1} = \frac{x^2}{x-1}$$

and multiplying:

$$x \left(\frac{x}{x-1} \right) = \frac{x^2}{x-1}$$

Thus, $\quad 7 + \tfrac{7}{6} = 7 \times \tfrac{7}{6} = \tfrac{49}{6}$

This relationship is derived from the fact that the square of any number is equal to unity plus the product of the preceding and succeeding numbers.

7. The number 123456789 (that is, containing all nine

digits in ascending order), if multiplied, gives the following results:

Multiply by 2		246913578
,, ,, 3		370370367
,, ,, 4		493827156
,, ,, 5		617283945
,, ,, 6		740740734
,, ,, 7		864197523
,, ,, 8		987654312

In each case, except where the multiplier is a multiple of 3, all the nine digits are repeated again in the result although in no consistent order. When the multiplier is a multiple of 3, the first three digits of the result are repeated to give the next three digits, whilst the same three digits minus the multiplier give the last three digits of the result. Similarly, related results introducing the zero sign in the answer, can be obtained by multiplying by some higher numbers.

8. If the number in the last paragraph is reversed, we obtain the number 987654321. Multiplication of this number gives the following pattern numbers:

Multiply by 2		1975308642
,, ,, 4		3950617284
,, ,, 5		4938271605
,, ,, 7		6913580247
,, ,, 8		7901234568

All the digits reappear with the addition of the zero sign in the results.

9. If the original two numbers in the last two paragraphs are written in the form of a subtraction, a new number, still containing all the digits, will appear.

$$987654321$$
$$123456789$$
$$\overline{864197532}$$

This number is very nearly the same as was obtained for the multiplier 7 in paragraph 7 above.

10. It is possible to find certain numbers of eight different digits which, when multiplied by the remaining ninth digit, will give results containing all nine digits:

$$51249876 \times 3 = 153749628$$
$$16583742 \times 9 = 149253678$$
$$32547891 \times 6 = 195287346$$

and also some variants:

(a) where a zero is introduced in the result:

$$675412398 \times 2 = 1350824796$$

(b) using a two-digit number as multiplier:

$$8745231 \times 96 = 839542176$$

(c) using only some of the digits:

$$8 \times 473 = 3784$$
$$15 \times 93 = 1395$$
$$35 \times 41 = 1435$$

(d) where the same number can be obtained by the multiplication of two different sets of numbers which are nevertheless composed of the same digits:

$$12 \times 483 = 42 \times 138 = 5796$$
$$18 \times 297 = 27 \times 198 = 5346$$

11. The number 142857 has the remarkable property that if it is multiplied by any number from 1 to 6, the results will always consist of the same digits in the same cyclic order, but commencing at different points.

$$142857 \times \quad 1 = 142857$$
$$2 = 285714$$
$$3 = 428571$$
$$4 = 571428$$
$$5 = 714285$$
$$6 = 857142$$

Multiplication by 7, however, gives a different result:

$$142857 \times \quad 7 = 999999$$

PERFECT NUMBERS AND SOME ODDITIES

Multiplication by the numbers 8 to 13 gives related results:

$$142857 \times 8 = 1142856$$
$$9 = 1285713$$
$$10 = 1428570$$
$$11 = 1571427$$
$$12 = 1714284$$
$$13 = 1857141$$

In each of these results there is, of course, an extra digit. Five of the six original digits appear in each result; the other two digits in each result, if added together, supply the missing sixth original digit.

12. It has been contended that the numbers 142857 and 285714 are the only numbers (lower than one million) which can be multiplied merely by removing the first digit and adding it on again at the end. This statement is not strictly correct. There are two exceptions, one of them fundamental.

(a) Every number consisting of x digits can be shown as a number of $x+1$ digits, where the first digit is a nought. The number 1234 is really the same as 01234. The nought in the position shown is always omitted in practice but it is always understood to be there. Its inclusion means precisely the same as its omission (that is, that there are no units of the fourth power of 10 in the composition of the number). The number 01234 is obviously multiplied by 10 when the nought is taken from the first position and added on in the end position.

(b) Apart from the foregoing, it may be true that the two numbers quoted are the only ones which can be multiplied by *a whole number* by moving the first digit to the other end. It is certainly not true if we include multiplication by quite simple fractions.

Thus:

$$153846 \times \tfrac{7}{2} = 538461$$

This new number (153846) is exactly twice 76923. If this be

expressed as 076923, it will be seen to have properties similar to those of 142857.

```
 1 ×076923 =076923
 2         =         153846
 3         =230769
 4         =307692
 5         =         384615
 6         =         461538
 7         =         538461
 8         =         615384
 9         =692307
10         =769230
11         =         846153
12         =923076
```

These results differ from those obtained in the multiplication of 142857 in that they give rise to numbers consisting of two distinct sets of digits. Each set of digits, however, is always repeated in the same cyclic order.

The numbers 142857 and 076923 have a common origin. They each represent a cycle of recurring decimals which result from the division of unity by a prime number (in these cases, the primes are 7 and 13 respectively). Recurring decimals are considered in more detail in the next chapter.

10

Recurring Decimals

In the previous chapter, attention was drawn to the peculiar properties of the number 142857. There are, in addition to this, many numbers—usually much greater—with similar or related properties and they are all of one type.

The number 142857 is not just a haphazard stringing together of digits but represents the cycle of recurring decimals which results from the division of unity by the number 7. Thus, $\frac{1}{7} = \cdot\dot{1}4285\dot{7}$ where the dots over the first and last digits mean that this set of six digits is repeated over and over again always in the same order. In the same way, the number 076923 (to which reference has also already been made) is derived from the decimal equivalent of the fraction $\frac{1}{13}(= \cdot\dot{0}7692\dot{3})$.

These sets of digits which recur over and over again are referred to here as recurring decimals, and these always result from the division of unity by a prime number (other than 2 or 5) or by any number containing such a prime number as one of its factors. The division of unity by 2 or 5 or any power of these numbers will not give rise to recurring decimals, but division by any number having 2 or 5 as one of its factors will nevertheless give recurring decimals if the divisor has any other prime as another factor.

In some cases digits recur immediately ($\frac{1}{3} = \cdot\dot{3}$) or following one other digit which is itself not repeated ($\frac{1}{6} = \cdot 1\dot{6}$). In most cases, however, there results a cycle of decimals in which a group of digits (as distinct from a single digit) recurs in the same order.

If the two numbers 142857 and 076923 are re-examined, the following two facts emerge:

(a) Eventually in the process of multiplication a number appears consisting entirely of a repetition of the digit 9.

(b) The digital root of each is 9.

(a) A little consideration makes it at once apparent that if the number derived from a cycle of recurring decimals is multiplied by the number by which unity was divided to obtain the original cycle, then a number consisting entirely of nines must result. Thus, the number 142857, which is derived from the division of unity by 7, becomes 999999 when multiplied by 7.

Here it is necessary to realize that $\frac{1}{7}$ does not equal ·142857. In fact it equals ·$\dot{1}4285\dot{7}$ carried on interminably. The cycle of digits in each case ends when the remainder from a division is 1, for that is, so to speak, where we came in. Since there must always be a remainder (unless we are dividing by 2 or 5) the process is obviously endless. But since also the cycle recommences on reaching the remainder 1, the number represented by the cycle digits does not result exactly from the division of 1 but results instead from the division of ·$\dot{9}$ (recurring to a limited extent, this extent varying for each divisor).

Thus, $\qquad \dfrac{1\cdot 0}{7} = \cdot 142857$ (recurring indefinitely)

but $\qquad \dfrac{\cdot 999999}{7} = \cdot 142857$ (exactly)

Therefore, in order to obtain the number 142857 we divide the number 999999. Consequently when 142857 is multiplied by 7, we return to the original number 999999.

(b) So far as the digital root of cyclic decimals is concerned, we may proceed as follows: From the foregoing it is seen that if we take as the divisor of unity the number x and obtain as the cyclic decimals the number y then $(x) \times (y)$ will give a number consisting entirely of nines. Now the digital root of any number consisting entirely of nines is itself 9, and it

RECURRING DECIMALS

therefore follows that the digital root of $(x)(y)$ must always be 9.

One of the following propositions therefore applies to each case:

(i) either x or y or both must have 9 as its digital root;
(ii) both x and y must have 3 or 6 as their roots;
(iii) either x or y must have 3 as its root, and the other must have 6 as its root.

From these propositions, the following may be derived:

(a) Unless the digital root of the divisor of unity is a multiple of 3, then the digital root of the cyclic number must be 9.
(b) If the digital root of the divisor is 3 or 6, the root of the cyclic number is 3, 6 or 9.
(c) If the digital root of the divisor is 9, the root of the cyclic number is 1, 3, 6 or 9.
(d) The digital root of a cyclic number can never be 2, 4, 5, 7 or 8.

Another way of 'building up' the cycle ·142857, and other similar cycles, has, at first sight, the appearance of magic. This is done by starting with the number 14 and then doubling in successive stages, the results being moved two places to the right at each stage.

```
       14
        28
         56
          112
           224
            448
             896
              1792
               3584
      ─────────────────
      142857142857142    etc.
```

Each of the above stages is shown as a multiplication, but instead really results from a division, using 7 as the divisor of unity. Here it is necessary to remember that instead of dividing into 1, we can divide into 10 or 100 and still get the same digits, merely moving the decimal point afterwards. We cannot divide 1 by 7 and so we normally bring down a nought so that, in fact, we are dividing 10 by 7. If, however, we extend the process and bring down two noughts together we are really dividing 100 by 7. The important point is that the resulting digits will be the same. A second point to consider is that every number will divide into a larger number an integral number of times and (unless it is also a factor of the larger number) leave a remainder. A third point is that, in dividing a large number involving many successive divisions, the remainder from each division is important as providing the basis for the next division.

We can now proceed as follows:

(*a*) We cannot divide 1 by 7, so we bring down two noughts.
(*b*) Divide 100 by 7. This equals 14, and remainder 2.
(*c*) Carry forward the remainder (2) and add two noughts. Divide 200 by 7. This equals 28, and remainder 4.

This process can be best seen as follows:

Operation	Result	Remainder
100 ÷ 7	14	2
200 ÷ 7	28	4
400 ÷ 7	56	8

It is at once apparent why, having obtained the result of the first division (giving 14) all that is necessary to obtain successive answers is to double the previous answer, because each successive remainder is double the previous remainder. Algebraically this can be expressed: if $\frac{100}{x}$ gives y and a remainder of 2, then $\frac{200}{x}$ will give $2y$ and a remainder of 4. Since the numerator is doubled, then both the result and the remainder are also doubled.

RECURRING DECIMALS

The same principle can be applied to other cycles. In the cycle for $\frac{1}{13}$ we have 076923. The division of 100 by 13 leaves a remainder of 9 and this gives the relationship between groups from which the cycle can be built up. Thus, instead of multiplying in stages by 2 (as in the case of 142857) we multiply by 9:

```
07
 63
  567
   5103
    45927
     413343
      3720087
       33480783
```

076923076   etc.

Similarly the principle can be extended if we divide 1000 instead of 100 and put the digits in groups of three. Dividing 1000 by 13 gives 76 and 12 remainder. The multiplier is therefore 12 and proceeding as before, multiplying successively by 12, we have

```
076
  912
   10944
     131328
```

07692307   etc.

There is yet another way in which the cycle 142857 can be built up, as follows:

7	7
7×50	350
7×50^2	17500
7×50^3	875000
7×50^4	43750000
7×50^5	2187500000
etc.	——142857

The cycle may therefore be represented as the sum of an infinite geometric progression whose first term is 7 and whose common ratio is 50. All recurring decimal cycles can be built up in the same way. Their first terms and common ratios are found thus. Each cycle is equivalent to a vulgar fraction (in the above case, $\frac{1}{7}$). Convert the fraction into another in which the last digit of the divisor is 9 (thus $\frac{1}{7}$ becomes $\frac{7}{49}$). The new numerator (7) is then the first term of the geometric progression, and the common ratio is the new divisor added to unity $(49+1=50)$.

Investigation of other cycles is most illuminating. A few examples follow. The results of various stages of multiplication are now given in such a way as to render comparison of digits easier.

Derivative of $\frac{1}{17} =$ ·0588235294117647 (Digital root $= 9$)

Multiply by 2 = ·1176470588235294
,, ,, 3 = ·1764705882352941
,, ,, 4 = ·2352941176470588
,, ,, 5 = ·2941176470588235
etc.

Derivative of $\frac{1}{19} =$ ·052631578947368421 (Digital root $= 9$)

Multiply by 2 = ·105263157894736842
,, ,, 3 = ·157894736842105263

In all the foregoing examples ($\frac{1}{7}$, $\frac{1}{13}$, $\frac{1}{17}$ and $\frac{1}{19}$) the derivatives have another feature in common. They all consist of an even number of digits. If these are separated into two groups by a line drawn between the two central digits, then each digit in the first group when added to the digit in the same position in the second group will give the number 9.

$$142857 \quad 142+857=999$$
$$076923 \quad 076+923=999$$

Not all derivatives have this property. The derivative of $\frac{1}{43}$, for example, cannot be multiplied by 2 or 3 merely by moving its digits (though it can be thus multiplied by 4 and some other numbers), nor can it be split into two groups so that relative pairs of digits sum to 9. It retains, however, the common property of having 9 as its digital root and, in addition, has another property. If it is split into two groups, the following results are obtained.

Derivative of $\frac{1}{43}$ = ·023255813953488372093
Split into two groups = 2325581395
  (ignoring the zero) = 3488372093
Divide first group by 2 = 1162790697
  (ignoring final remainder)
  ,, second group by 3 = 1162790697
  (ignoring final remainder)

This property is not uncommon, although it may take slightly different forms for different derivatives. The derivative for $\frac{1}{71}$ is a particularly good example, and is therefore shown in full in the table on page 104.

It will be seen that, no matter at what digit we start, this digit, taken with the next six or seven digits (as required) in rotation will give a multiple or near-multiple of 140845.

The usual method of ascertaining the digits appearing in a cycle of decimals is by ordinary division. These digits can, however, be found by a reverse method, whereby it is possible to find the last digit in a cycle without knowing the first digit. To understand this method, it must be remembered that when obtaining a complete cycle of digits, we are really dividing a number consisting entirely of a repetition of the digit 9. When dividing by 7, for instance, it is obvious that we shall obtain a cycle which when multiplied by 7, will give a number consisting entirely of nines and this number will also be exactly divisible by 7. It follows therefore that the last digit of the cycle must be such as will, when multiplied by 7, give a number of two digits the second digit of which must be 9. The only number which satisfies this requirement

Derivative of $\frac{1}{71}$ =	·01408450704225352112676056338028169	*Multiple*
Split into groups:	140845	1
	281690	2
	422535	3
	563380	4
	704225	5
	845070	6
	1126760	8
	1267605	9
	1408450	10
	1690140	12
	2112676	15
	2253521	16
	2535211	18
	2676056	19
	3380281	24
	3521126	25
	3802816	27
	4084507	29
	4507042	32
	5211267	37
	5352112	38
	5070422	36
	6338028	45
	6760563	48
	9014084	64
	6056338	43
	6901408	49
	7605633	54
	8028169	57
	8169014	58

is 49 ($=7\times 7$) and the last digit of the cycle must therefore be 7.

But the digit 4 in the number 49 is a remainder from a previous division in which another digit 9 has been 'brought down'. As the remainder from 9 is 4, it follows that the next to the last digit in the cycle must be such that, when multi-

RECURRING DECIMALS

plied by 7, gives a number ending in the digit 5. This number can only be 5. We now have:

```
7 ) 999,999 ( ····57
     ··
     ·9
     ··
      ·9
      .6
      ──
       39
       35
       ──
        49
        49
        ··
```

The digit 3 in the number 35 is then taken (as shown) as the previous remainder. 3 deducted from 9 leaves 6. The next digit in the cycle, still working backwards, must therefore be 8, since $8 \times 7 = 56$, and this is the only two-digit multiple of 7 which has 6 as its second digit.

Another method in which we may calculate the digits in a cycle in a backwards order is as follows. The division from which the cycle is to result is first shown in the form of a fraction (thus, $\frac{1}{7}$). Both the numerator and the divisor are then multiplied by the smallest number which will make the divisor a number whose last digit is 9. (Thus, $\frac{1}{7} \times 7 = \frac{7}{49}$.) Then the last digit of the cycle will be the same as the new numerator (that is, in our example, 7). The other digits are then found in a different way. First we increase the new divisor by 1 so as to give a multiple of 10 (that is, $49+1=50$). This multiple of 10 is then divided by 10 or, in other words, the zero sign in the end position is crossed out. The number which finally emerges from these operations is then taken as what we shall call the Relative of the cyclic recurrence. In the case of $\frac{1}{7}$, the Relative is 5.

Next, two columns are constructed and numbers are entered as explained at each step:

		Digit Column	Carry-forward Column
(a)	The last digit of the cycle is known	7	
(b)	Multiply this digit (7) by the Relative (5) =35. Write the digit 5 in the digit column and the digit 3 in the carry-forward column	5	3
(c)	Multiply the last digit in the digit column by the Relative and add in the carry-forward. $5 \times 5 + 3 = 28$. Write the digit 8 in the digit column and 2 in the carry-forward column	8	2
(d)	Proceed as in (c), thus: $5 \times 8 + 2 = 42$	2	4
(e)	$5 \times 2 + 4 = 14$	4	1
(f)	$5 \times 4 + 1 = 21$	1	2
(g)	$5 \times 1 + 2 = 7$	7	0

This process is continuous, and the digits of the cycle can be found in reverse order in the digit column.

One of the basic points outstanding is the question of why decimals recur at all, but this is easily shown. We have already seen that when unity is divided by a number larger than itself, we have to 'bring down' a succession of noughts to give a decimal result. We are dividing 10 or 100 or any other power of 10 (according to the number of noughts employed) and then shifting the decimal point. If the divisor is a factor of any power of 10, then the decimal will have a definite period and the digits will not recur. If, however, the divisor is not such a factor then it will not divide into 10 (or a power thereof) an exact integral number of times and there will always be a remainder. There can, therefore, be no end to the decimal expression.

But the number of *different* remainders possible is limited by relation to the divisor. When we divide by 7, the only possible remainders are 1, 2, 3, 4, 5 and 6. This clearly applies to any divisor, so that if we divide by the number N, then there are only $N-1$ possible remainders. When all these different remainders have appeared once in the course of the division, the stage will be reached when one of them is repeated and at that point the whole process begins again.

The number of digits in each cycle is also related to the expression $N-1$, in that the number of digits is always equivalent to $N-1$ or a factor thereof, if N is prime (other than 2 or 5).

II

Congruences

Every integer may be expressed as the multiple of a lower integer plus a remainder. The number n can accordingly be expressed as

$$qm+r$$

where r is the remainder when n is divided by m; qm being a multiple of m. Where n is expressed as a multiple of m, and m is a factor of n, the remainder will obviously be equal to nought.

The number n will not, however, be the only number which gives the remainder r when divided by m, and this is the basic fact upon which the theory of congruences was founded by the mathematician Gauss.

The numbers 40 and 64, for instance, give the same remainder of 4 when divided by 12. This is another way of saying that both numbers are equal to multiples of 12 added to 4. They both, therefore, bear a certain common relationship in their attitude to the number 12. In mathematical language the latter number is called the 'modulus' in respect of which the two other numbers have similar properties, and these two numbers are said to be 'congruent' to each other for that modulus.

Thus, if $$a = qm+r$$
and $$b = qm+r$$

then a is said to be congruent to b for the modulus m, and the relationship is expressed as:

$$a \equiv b \pmod{m}$$

The quantities a and b are termed residues of each other for

CONGRUENCES

the given modulus, and the totality of all integers congruent to each other for a particular modulus are said to comprise the residue class for that modulus.

From the definition, two numbers are congruent to each other for a specified modulus when their remainders are identical, and it therefore follows that the difference between the two congruent numbers must be exactly divisible by the modulus. That is, if $a \equiv b \pmod{m}$ then $(a-b)$ must be divisible by m.

Another development of the congruence $a \equiv b \pmod{m}$ is that $a^n \equiv b^n \pmod{m}$.

Thus, $\qquad 4 \equiv 7 \pmod{3}$

and $\qquad 4^2 \equiv 7^2 \pmod{3}$

as will be seen by inspection. This further congruence arises because:

$\qquad 4 = 3+1$
and $\qquad 7 = (2)(3)+1$
so $\qquad 4^2 = (3+1)^2 = 3^2 + (2)(3) + 1$
and $\qquad 7^2 = (6+1)^2 = 6^2 + (2)(6) + 1$

and both of these extensions clearly differ by 1 from multiples of 3.

This fact enables us to ascertain by a very short process what remainder will be left when a number, expressed as a power of a smaller number, is divided by a third number. There is not even any need to calculate the full digital representation of the original number. As an example we show the procedure for finding the remainder when 2^{28} is divided by 15.

$\qquad 2^4 = 16$ and $16 \equiv 1 \pmod{15}$
therefore $\qquad 2^4 \equiv 1 \pmod{15}$
and $\qquad (2^4)^7 \equiv 1^7 \pmod{15}$
so $\qquad 2^{28} \equiv 1 \pmod{15}$

and the remainder will therefore be 1.

An important use of the congruence concept has already been shown in Chapter 6, for the tests of divisibility included are all based on the conversion of certain numbers into the form of multiples of other numbers plus certain remainders. Congruences also assist in certain other cases of factorization. We can, for example, prove that the number 23 is a factor of the expression $2^{11}-1$ as follows:

$$2^5 = 32 \equiv 9 \pmod{23}$$

Therefore, by squaring each side of the congruence,

$$2^{10} \equiv 81 \pmod{23}$$
$$\equiv 12 \pmod{23}$$

(12 being the new remainder when 81 is divided by 23, so that $81 \equiv 12 \pmod{23}$.

But $\qquad 2 \equiv 2 \pmod{23}$
so $\qquad 2^{10} \times 2 \equiv 12 \times 2 \pmod{23}$
or $\qquad 2^{11} \equiv 24 \pmod{23}$
$\qquad \equiv 1 \pmod{23}$
since $\qquad 24 \equiv 1 \pmod{23}$

Therefore, when $2^{11}-1$ is divided by 23 there is no remainder, and accordingly 23 must be a factor of $2^{11}-1$.

When dealing with residue classes to the same modulus, the multiplication of two terms involves the multiplication of their respective remainders.

$$16 \equiv 2 \pmod{14}$$
and $\qquad 18 \equiv 4 \pmod{14}$
so $\qquad 16 \times 18 \equiv 2 \times 4 \pmod{14}$
$\qquad \equiv 8 \pmod{14}$

It is important to note, however, that the reverse procedure is not necessarily correct for all congruences. Whilst the element $8 \pmod{14}$ can, by reversing the above procedure, be split into two elements $2 \pmod{14}$ and $4 \pmod{14}$, no similar procedure can be employed in all cases.

The number 16 is, for instance, congruent to 6 for the modulus 10, but it is not possible to split the element 6 into two factors so as to give two relative residues whose multiplication would give a result congruent to $6 \pmod{10}$. If we take the factors 3 and 2 of the element 6, one of the residues must be a multiple of 10, plus 3; the other must be a multiple of 10, plus 2. If these two residues are multiplied we would have

$$(10_1+3)(10_2+2)$$

but (10_1+3) would be an odd number, so that if $16 \equiv (10_1+3)(10_2+2)$, then 16 would be divisible by an odd number and this is ridiculous.

The theory of congruences is greatly to the fore in general number theory and has been called into use in many of the problems involved in the discovery of tests for primality. As an example, we have Wilson's theorem that if p is prime then $(p-1)!+1 \equiv 0 \pmod{p}$; or conversely that if $(p-1)!+1 \equiv 0 \pmod{p}$ then p is prime.

As an example, we take the prime value $p = 11$.

Then $\quad (p-1)!+1 = 10!+1 = 3628801$

This number is divisible by 11, so

$$3628801 \equiv 0 \pmod{11}$$
and $\quad (p-1)!+1 \equiv 0 \pmod{11}$

Although this affords a test for primality, the very large numbers involved are so great as to render it unpractical. Adaptations of other congruence theorems, mainly the result of the work of the French mathematician Lucas, give a test for primality but have the failing that they do not reveal the actual factors of numbers shown to be composite.

12

Irrationals, Imaginary Numbers and Continued Fractions

Whereas the exact nature of many number relationships may be more easily demonstrated by constructional representation, the proportionate dimensions resulting from certain other constructions served only to create confusion in the early days of Mathematics. The numbers thus produced were not of the ordinary kind. They were not integral and, indeed, were subsequently proved to be incapable of expression in ordinary notation at all.

The equation: $a^2 = b^2 + c^2$, when applied to a right-angled triangle where $b = c = 1$, produces the result:

$$a = \sqrt{2}$$

The inability of the Pythagoreans to represent $\sqrt{2}$ as an integer was a grave eruption in their philosophical world.

Another inexpressible quantity which has exercised the minds of mathematicians for centuries is the ratio between the diameter and the circumference of a circle. If the diameter is measured off along the circumference, it is found that the latter is slightly more than three times as great as the former. This ratio is represented by the symbol π (the Greek letter, Pi), and one of the great mathematical pastimes of the past was the attempt to 'square the circle', until it was discovered that π was inexpressible, or transcendental, and that the objective was therefore impossible.

Euclid proved the impossibility of expressing $\sqrt{2}$ as a rational number; his reasoning being as follows. If such a

IRRATIONALS AND IMAGINARY NUMBERS

rational number could exist, then it could be represented alternatively as:

$$\frac{a}{b}$$

where the latter fraction is in its lowest terms—that is, all common factors have been cancelled out. It follows that either a or b, or both, must be odd; for if both a and b are even, then the fraction cannot be in its lowest terms.

If $$\frac{a}{b} = \sqrt{2}$$

then $$\frac{a^2}{b^2} = 2$$

and $$a^2 = 2b^2$$

This means that a^2, and therefore a as well, is even. But if a is even it is a multiple of 2 and may be represented alternatively as $2m$.

Then $$a^2 = 4m^2$$
but $$a^2 = 2b^2$$
so $$2b^2 = 4m^2$$
and $$b^2 = 2m^2$$

so that b^2, and therefore b as well, is even.

We therefore find that both a and b are even, and this means that $\frac{a}{b}$ is not in its lowest terms. This contradicts the original assumption and proves that $\sqrt{2}$ cannot be represented rationally.

Although irrationals cannot be calculated exactly, it is possible to obtain approximate values which are quite adequate for practical requirements. Various mathematicians have calculated the value of π to remarkable lengths, but such values as have been produced have no practical use, for after the first few significant figures, subsequent digits provide a degree of accuracy which is never called into use.

Archimedes attacked the problem of calculating the value of π by the method of approximating to a circle by means of alternately inscribed and circumscribed polygons. This, of course, can only give an approximate value because, no matter how near a polygon may be made to resemble a circle, it can never actually become a circle. The polygons may therefore be said to converge on the circle without actually approaching it, and the unattainable (i.e. the circle) may be said to constitute the limit of such a convergence.

This concept of a convergence to a limiting value is of immense importance in theories concerning the irrational numbers and the infinite nature of their composition.

One method of calculating approximate values is by the use of continued fractions. The square root of 2 may be represented by this means as follows:

$$\sqrt{2} = 1 + \cfrac{1}{2 + \cfrac{1}{2 + \cfrac{1}{2 + 1 \ldots \text{etc.}}}}$$

In order to represent $\sqrt{2}$ in this way, we first extract the integral part, 1, and express the remainder as a fraction:

$$\sqrt{2} = 1 + \frac{1}{x_1}$$

where x_1 is greater than 1. From this,

$$\sqrt{2} - 1 = \frac{1}{x_1}$$

so

$$\frac{1}{\sqrt{2}-1} = x_1$$

But

$$\frac{1}{\sqrt{2}-1} = \sqrt{2}+1; \quad \text{because } (\sqrt{2}-1)(\sqrt{2}+1) = 1$$

therefore

$$x_1 = \sqrt{2} + 1.$$

The integral part of x_1 is therefore 2. We extract this from x_1 and express the remainder as a fraction:

$$x_1 = 2 + \frac{1}{x_2}$$

From this $\quad x_2 = \dfrac{1}{x_1 - 2}$

but $\quad x_1 = \sqrt{2} + 1$

so $\quad \dfrac{1}{x_1 - 2} = \dfrac{1}{\sqrt{2} + 1 - 2} = \dfrac{1}{\sqrt{2} - 1}$

and $\quad x_2 = \dfrac{1}{\sqrt{2} - 1} = \sqrt{2} + 1$

Thus, x_1 and x_2 are identical, as indeed will be x_3, x_4, etc. The integral part of each is 2 and we therefore have a continued fraction which consists of partial denominators all of which are 2.

An easier way of writing out a continued fraction is as follows:

$$\sqrt{2} = 1 + \frac{1}{2} + \frac{1}{2} + \frac{1}{2} + \frac{1}{2} \quad \text{etc.}$$

The plus signs, printed in line with the divisors instead of in the usual position, indicate that the expression is a representation of a continued fraction. There is, therefore, a great and fundamental difference between the expressions:

(a) $\dfrac{1}{2} + \dfrac{1}{2} + \dfrac{1}{2} + \dfrac{1}{2} + \dfrac{1}{2}$

(b) $\dfrac{1}{2} + \dfrac{1}{2} + \dfrac{1}{2} + \dfrac{1}{2} + \dfrac{1}{2}$

Any positive number may be represented as a continued fraction. The general procedure is now detailed. We first extract the integral part, if any, leaving a fractional part to convert. The denominator is then divided by the numerator, giving an integral answer plus a remainder. The remainder is then used as a new divisor to divide into the previous divisor, and this process is continued as far as required, a note being made of the integer derived from each successive division. These integers form the partial denominators of the continued fraction.

It will be seen that this procedure is much the same as for the calculation of the highest common factor of two numbers. The fraction $\frac{521}{1023}$ is converted to a continued fraction as shown:

```
521 ) 1023 ( 1
      521
      502 ) 521 ( 1
            502
             19 ) 502 ( 26
                  494
                    8 ) 19 ( 2
                        16
                         3 ) 8 ( 2
                             6
                             2 ) 3 ( 1
                                 2
                                 1 ) 2 ( 2
```

and is therefore equivalent to:

$$\frac{1}{1+}\frac{1}{1+}\frac{1}{26+}\frac{1}{2+}\frac{1}{2+}\frac{1}{1+}\frac{1}{2}$$

This is a continued fraction for a rational expression and it has a definite ending. A continued fraction for an irrational expression is endless, and it depends upon the degree of accuracy required as to how many terms are calculated. As

IRRATIONALS AND IMAGINARY NUMBERS 117

we proceed, the approximation becomes closer to the actual value at each stage, and the different values thus obtained are called the convergents of the fraction. Successive convergents of the continued fraction for $\frac{521}{1023}$ are:

1st convergent: $\frac{1}{1}$

2nd   ,, $\frac{1}{2}$

3rd   ,, $\frac{27}{53}$

4th   ,, $\frac{55}{108}$

5th   ,, $\frac{137}{269}$

6th   ,, $\frac{192}{377}$

7th   ,, $\frac{521}{1023}$

These convergent values are calculated direct from the continued fraction, taking one further step for each convergent. After the first two convergents have been calculated, subsequent convergents may be calculated from the two preceding convergents. This may be shown by substituting letters for the respective integers. For example:

1st convergent: $\frac{1}{1} = \frac{a}{b}$

2nd   ,, $\frac{1}{2} = \frac{c}{d}$

3rd term in fraction: $\frac{1}{26} = \frac{e}{f}$

Then the third convergent may be shown as related to the values a, b, c, d, e and f as follows:

$$\text{3rd convergent} = \frac{(c \times f) + (a \times e)}{(d \times f) + (b \times e)}$$

$$= \frac{(1 \times 26) + (1 \times 1)}{(2 \times 26) + (1 \times 1)} = \frac{27}{53}$$

The simplest irrational continued fraction of all is:

$$\frac{1}{1+}\frac{1}{1+}\frac{1}{1+}\frac{1}{1+}\frac{1}{1+} \text{ etc.}$$

or

$$\cfrac{1}{1+\cfrac{1}{1+\cfrac{1}{1+\cfrac{1}{1+1}}}} \text{ etc.}$$

and the convergents of this fraction are:

$$\tfrac{1}{1}, \tfrac{1}{2}, \tfrac{2}{3}, \tfrac{3}{5}, \tfrac{5}{8}, \tfrac{8}{13} \text{ etc.}$$

Both the numerators and the denominators are derived from the series of Fibonacci numbers, each of which is equal to the sum of the two preceding numbers:

$$1 \quad 1 \quad 2 \quad 3 \quad 5 \quad 8 \quad 13 \quad 21 \quad 34 \quad 55 \quad 89 \text{ etc.}$$

The convergents of continued fractions are alternately greater than and less than the true value, the error growing smaller either way as further convergents are reached.

The use of continued fractions also gives a solution to any equation of the form:

$$ax - by = 1$$

where a and b are relatively prime positive integers. To solve the equation:

$$32x - 51y = 1$$

we set up the continued fraction for $\tfrac{32}{51}$ and expand it as follows:

$$\cfrac{1}{1+\cfrac{1}{1+\cfrac{1}{1+\cfrac{1}{2+\cfrac{1}{6}}}}}$$

IRRATIONALS AND IMAGINARY NUMBERS

The convergent values for this expression are, in order:

$$\tfrac{1}{1}, \ \tfrac{1}{2}, \ \tfrac{2}{3}, \ \tfrac{5}{8}, \ \tfrac{32}{51}$$

The values of x and y which will satisfy the equation are then found in the last convergent but one, if the total number of convergents is odd (as in this case). Thus $x=8$; and $y=5$, these values being proved as under:

$$(8 \times 32) - (51 \times 5) = 1$$

The required values are found in a slightly different way if the total number of convergents is even. For example, to solve the equation:

$$13x - 48y = 1$$

we set up the fraction $\tfrac{13}{48}$ and expand it to:

$$\cfrac{1}{3 + \cfrac{1}{1 + \cfrac{1}{2 + \cfrac{1}{4}}}}$$

and the convergents are:

$$\tfrac{1}{3}, \ \tfrac{1}{4}, \ \tfrac{3}{11}, \ \tfrac{13}{48}$$

If we substitute the integers in the last convergent but one (i.e. 3 and 11) for values of y and x, we obtain the result:

$$(13 \times 11) - (48 \times 3) = -1$$

whereas we require values which will give a positive result instead of a negative one.

We can, however, find the required values by deducting the integers in the third convergent from the values of a and b in the equation

$$ax - by = 1$$

If we call the integers in the third convergent x_1 and y_1 to distinguish them from the true values of x and y, then

$$x = b - x_1 = 48 - 11 = 37$$
$$y = a - y_1 = 13 - 3 = 10$$

and these adjusted values will be found to satisfy the equation

$$(13 \times 37) - (10 \times 48) = 1$$

Various approximate values for the more important of the irrational numbers have been suggested at different stages of mathematical development. The square root of 2 has an approximate decimal equivalent:

$$\sqrt{2} = 1 \cdot 41421356 \ldots$$

A simpler, though less accurate, approximation is $\frac{17}{12}$ for

$$(\sqrt{2})^2 = 2$$

and

$$\left(\frac{17}{12}\right)^2 = \frac{289}{144} = 2 \text{ approx.}$$

The first few convergents derived from the continued fraction for $\sqrt{2}$ are:

$$\frac{1}{1} \quad \frac{3}{2} \quad \frac{7}{5} \quad \frac{17}{12} \quad \frac{41}{29}$$

An approximate value for π is $3 \cdot 14159265 \ldots$, but several other useful approximations, suitable for the purposes for which they were suggested, have been used:

(a) $3\frac{13}{81}$ (derived from $(\frac{16}{9})^2$)

(b) $\sqrt{10}$

(c) Between $3\frac{1}{7}$ and $3\frac{10}{71}$

The first terms of the continued fraction for π are:

$$3 + \cfrac{1}{7 + \cfrac{1}{15 + \cfrac{1}{1 + \cfrac{1}{292 + \ldots}}}}$$

and successive convergents are:

(i) 3
(ii) $\frac{22}{7} = 3 \cdot 142857$ approx.
(iii) $\frac{333}{106} = 3 \cdot 141509$,,
(iv) $\frac{355}{113} = 3 \cdot 14159292$,,

IRRATIONALS AND IMAGINARY NUMBERS

In addition to the foregoing, there have been a surprising number of suggestions as to how π could be expressed:

$$\pi = 2\left[\frac{2 \cdot 2}{1 \cdot 3} \times \frac{4 \cdot 4}{3 \cdot 5} \times \frac{6 \cdot 6}{5 \cdot 7} \times \frac{8 \cdot 8}{7 \cdot 9} \times \cdots\right]$$

$$= 4\left[1 - \frac{1}{3} + \frac{1}{5} - \frac{1}{7} + \frac{1}{9} - \frac{1}{11} + \frac{1}{13} - \cdots\right]$$

$$= 2\left[1 + \left(\frac{1}{3}\right)\left(\frac{1}{2}\right) + \left(\frac{1}{5}\right)\left(\frac{1 \cdot 3}{2 \cdot 4}\right) + \left(\frac{1}{7}\right)\left(\frac{1 \cdot 3 \cdot 5}{2 \cdot 4 \cdot 6}\right) \cdots\right]$$

$$= 4\left[\cfrac{1}{1 + \cfrac{1^2}{2 + \cfrac{3^2}{2 + \cfrac{5^2}{2 + 7^2 \cdots}}}}\right]$$

Another remarkable quantity is the number e, the approximate value of which is 2·718281828459, so named after its discoverer, Leonhard Euler. This number has many applications. It is closely associated with statistical theory and is of immense importance in the calculus. It is the base to which natural logarithms are calculated, and is also concerned in compound interest calculations. This last use helps to give a clearer picture of the nature of the number e and is therefore well worth a little attention.

If £1 is loaned at 2 per cent. compound interest per annum, at the end of the first year the debt will have increased to

$$£(1 + \cdot 02)$$

If, however, the interest were to be reckoned at the end of every month instead of at the end of every year, it would be found to increase more rapidly. At the end of the first month, the total debt would be:

$$£\left(1 + \frac{\cdot 02}{12}\right)$$

and at the end of the year:

$$£\left(1+\frac{\cdot 02}{12}\right)^{12}$$

If these expressions are simplified, we arrive at a different value at the year end for each of the different methods employed.

By calculating interest still more frequently (e.g. weekly), we find that the value at the end of the year is again higher than if we calculate yearly or monthly. It does not, however, increase indefinitely. There is a value beyond which it will not increase, or in other words, there is a limit to its possible value. If the loan had been at 100 per cent. interest this limit would have been £2·718 approximately and its connexion with the number e is at once apparent. In the present example the limit is £2·718$^{(\cdot 02)}$ or £1·021.

The number e may be represented in a number of different ways:

$$e = 2 + \cfrac{1}{1 + \cfrac{1}{2 + \cfrac{2}{3 + \cfrac{3}{4 + 4 \ldots}}}}$$

$$= 1 + \frac{1}{1} + \frac{1}{1 \cdot 2} + \frac{1}{1 \cdot 2 \cdot 3} + \frac{1}{1 \cdot 2 \cdot 3 \cdot 4} \ldots$$

$$= 2 + \left[\frac{1}{1+2} + \frac{1}{+1} + \frac{1}{+1} + \frac{1}{+4} \ldots\right]$$

= the limit of the sequence: $(1\frac{1}{2})^2$, $(1\frac{1}{3})^3$, $(1\frac{1}{4})^4$, $(1\frac{1}{5})^5 \ldots$

In addition to irrational numbers of the type previously mentioned, there are other numbers, the values of which we cannot even calculate approximately. These are called *imaginary* numbers and the most important of these is i, equivalent to the square root of minus 1.

$$i = \sqrt{-1}$$

It is not possible to give any short explanation of what the square root of minus 1 may be said to represent, although it occurs frequently in calculations involved in wireless and electricity theories. A full justification of its uses may be found in advanced mathematical treatises, but it is of interest to note that its practical use is made possible by the fact, when it does appear in equations, it is usually cancelled out before the final answer is reached.

The number i has peculiar properties of its own, the most obvious of these being that successive powers of it are repetitive:

$$i = \sqrt{-1} \qquad\qquad i^5 = \sqrt{-1}$$
$$i^2 = -1 \qquad\qquad i^6 = -1$$
$$i^3 = -\sqrt{-1} \qquad\qquad i^7 = -\sqrt{-1}$$
$$i^4 = +1 \qquad\qquad i^8 = +1$$

This produces the relationship:

$$i^3 = -i$$

The numbers π, e and i may be related to each other by the remarkable equation:

$$e^{i\pi} = -1$$

whilst the two numbers e and π are featured in Stirling's Theorem, which states that:

$$\sqrt{(2\pi n)} \cdot e^{-n} n^n = n! \text{ approx.}$$

The introduction of imaginary numbers can have a remarkable effect upon real numbers. If any integral number is split into two integral parts and these two parts are multiplied together, the highest quotient that can result is the square of half the original number. If the original number be x, then the limit for possible quotients is $= \left(\dfrac{x}{2}\right)^2$.

The number 14, for example, may be split into pairs of

numbers and the multiplication of these parts may be shown as follows:

$14 = 1 + 13$	$1 \times 13 = 13$
$= 2 + 12$	$2 \times 12 = 24$
$= 3 + 11$	$3 \times 11 = 33$
$= 4 + 10$	$4 \times 10 = 40$
$= 5 + 9$	$5 \times 9 = 45$
$= 6 + 8$	$6 \times 8 = 48$
$= 7 + 7$	$7 \times 7 = 49$

No higher quotient is possible since the next stage is 8×6, and this is the same as the sixth stage shown above. The number 14 cannot, therefore, be split into two integral parts so that the multiplication of these parts results in a number greater than 49.

If we introduce imaginary numbers, however, the number 14 can be split into two parts to give any result we like:

If
$$7 + 7 = 14$$
Then
$$(7 + \sqrt{-10}) + (7 - \sqrt{-10}) = 14$$

for we have added $\sqrt{-10}$ and taken it away again. And,

$$(7 + \sqrt{-10}) \text{ multiplied by } (7 - \sqrt{-10}) = 49 - \sqrt{(-10)^2}$$
$$= 49 \pm 10$$
$$= 59 \text{ or } 39$$

13

Pseudo-Telepathy

A knowledge of number relationships enables the possessor of the knowledge to perform a number of tricks which, to the uninitiated, would appear to be based on thought transference or other magical powers. This knowledge enables us not only to understand certain of the more generally accepted party tricks but also, once the principles are thoroughly understood, to evolve new ones.

The following example, employing only elementary principles, is well known. For convenience, the process is tabulated:

(a) Take any number of three digits in which the difference between the first and last digits exceeds unity.
(b) Reverse the digits, thus obtaining another number.
(c) Deduct the smaller number from the larger and make a note of the result, this being the third number.
(d) Reverse the digits of the third number, giving a fourth number.
(e) Add the third and fourth numbers together. Their sum will always be 1089.

Examples:

	(i)	(ii)
	721	635
	127	536
	594	099
	495	990
	1089	1089

From the second example, it will be seen that the zero sign must always be used to fill any gap and must not merely be 'understood' to be there.

The fact that the number 1089 will always result from the above operations may easily be understood if it be remembered that any number '*abc*' is really a number of the form $100a+10b+c$. Every number of three digits is essentially of this form and the process may therefore be represented as follows, stage by stage, irrespective of the numerical values of the digits a, b and c.

Original number

	$100a$	$+\ 10b+c$
Reverse	$100c$	$+\ 10b+a$
Subtract	$100(a-c-1)$	$+\ 90\ +(10+c-a)$
Reverse	$100(10+c-a)$	$+\ 90\ +(a-c-1)$
Add	$100(-1+10)$	$+180\ +(10-1)$
$=$	900	$+180\ +\ \ \ \ 9$
$=1089$		

Since the first number is greater than its reversal, it follows that the digit a is greater than c, so that in order to subtract the former from the latter, it is necessary to borrow 10 from the tens column. As, however, there are no tens from which to borrow (because $10b-10b=0$), we have to borrow 100 from the hundreds column. Of this 100, we carry 10 to the units column to assist the subtraction of a from c, and the balance of 90 is then left in the tens column.

A similar proposition is to write down a sum of pounds, shillings and pence, so that the number of pounds is less than 12 but exceeds the number of pence by two or more, and then to treat this in exactly the same way as the number '*abc*' in the previous example.

Original	£8	15	6
Reverse	6	15	8
Subtract	1	19	10
Reverse	10	19	1
Add	12	18	11

The result is always the same. The proof is similar, it being again remembered that it is necessary to borrow when subtracting.

Original	x	z	y
Reverse	y	z	x
Subtract	$(x-y-1)$	$(19+z-z)$	$(y+12-x)$
Reverse	$(y+12-x)$	$(19+z-z)$	$(x-y-1)$
Add	$(-1+12)$	(38)	$(12-1)$
$=$	11	38	11
$=$	12	18	11

It is worthy of note that in both the above examples the third and fourth numbers are always multiples of 99 (that is, one less than 100), while the third and fourth sums of money are always multiples of 19*s*. 11*d*. (that is, one penny less than £1).

There are many ways of ascertaining the exact nature of a number selected by another person by instructing him to subject the number to various mathematical processes. A few examples follow. They are usually clothed in mystical significance by the performer telling the other person to select a particular number known only to himself, such as his age in years or his telephone number or the number of his house. Here they are shown in tabulated form, being a summary of the instructions issued by the performer.

EXAMPLE I

(*a*) When the number is selected, treble it.
(*b*) Divide the result by 2 (adding ½ to the result only if this is necessary to make it a whole number).
(*c*) Treble the result in (*b*).
(*d*) Divide by 9. This will give a certain result, either being a whole number only or being a whole number and a fraction. The fraction, if any, is ignored. We are interested only in the whole number, which the selector is asked to reveal and which we shall show

here as x. The original number is then known to be twice the final result (that is, $2x$) unless at the second stage it was necessary to add $\frac{1}{2}$ in which case the original number is equivalent to $2x+1$. The proof is as follows.

(i) If the original number is even, it is of the form $2x$ and the various stages may therefore be shown thus:

(a) $2x \times 3 \quad = 6x$
(b) $6x \div 2 \quad = 3x$
(c) $3x \times 3 \quad = 9x$
(d) $9x \div 9 \quad = x$

(ii) If the original number is odd, it is of the form $2x+1$.

(a) $(2x+1) \times 3 \quad = 6x+3$
(b) $(6x+3) \div 2 \quad = 3x+1\frac{1}{2}$
  add $\frac{1}{2} \quad = 3x+2$
(c) $(3x+2) \times 3 \quad = 9x+6$
(d) $(9x+6) \div 9 \quad = x+\frac{2}{3}$

EXAMPLE 2

(a) Select a number and multiply it by 5.
(b) Add 6.
(c) Multiply by 4.
(d) Add 9.
(e) Multiply by 5.

The result of these operations is made known; the original number is easily found by deducting 165 from the result, giving a number ending in two noughts. The digits to the left of these noughts form the original number. In other words, the result minus 165 is equivalent to one hundred times the original number.

Thus:

Selected number:	x
Multiply by 5:	$5x$
Add 6:	$5x+6$
Multiply by 4:	$20x+24$
Add 9:	$20x+33$

Multiply by 5: $100x+165$
Deduct 165: $100x$

EXAMPLE 3

In this example two numbers are selected, one person selecting an odd number and another person selecting an even number. The task of the third person is not to discover the actual numbers selected but, instead, to discover which person selected the even or the odd number. This is quite simple. The first person (A) should be told to multiply his number by any even number, and the second person (B) should be told to multiply his number by any odd number. They should then add the two results together and disclose the sum total.

If the total is odd, the A must have selected the even number, but if the total is even then B must have selected the even number.

The stages are as follows (assuming that A multiplies his number by 2, while B multiplies by 3).

(i) $2A$
(ii) $3B$
(iii) $2A+3B$

The sum total is therefore $2A+3B$. This can be even only if B is even; and so A will be odd. It can be odd only if B is odd and A will be even.

EXAMPLE 4

It is possible to discover three different selected numbers, less than 10, by the following process.

(i)   Three numbers are selected: x, y and z
(ii) Multiply one of them by 2 $=2x$
(iii) Add 3 $=2x+3$
(iv) Multiply by 5, and add 7 $=10x+22$
(v)   Add in the second number $=10x+22+y$
(vi) Multiply by 2, and add 3 $=20x+47+2y$
(vii) Multiply by 5, and add in the
      third number $=100x+235+10y+z$

When the final result is disclosed, deduct 235 leaving a balance of $100x+10y+z$. This is the form of all three-digit integral numbers and the digits of the final number are therefore the same as the original selections. Thus, if the numbers 3, 4 and 5 are selected, the respective stages are: (ii) 6; (iii) 9; (iv) 52; (v) 56; (vi) 115; (vii) 580; and $580-235=345$.

EXAMPLE 5

This is an example of how simple numerical relationships can be incorporated into party tricks so dressed up as to appear to remove the tricks from the realms of pure arithmetic. Someone is asked to open a book at any page and to choose any word in any of the first nine lines on the page, to make a note of that word and to close the book.

He is then told to:

(a) Double the page number and multiply by 5.
(b) Add 25.
(c) Add the line number and multiply by 10.
(d) Add the number of the word in the line.
(e) Subtract 250.
(f) Disclose the result.

This result will consist of a group of digits in which the number of the page, the number of the line, and the number of the word in that line all appear and in the order shown. Thus, if the second word in the third line of the ninth page is chosen, we finally obtain the result 932.

If the number of the page be P; of the line be L; and of the word be W; then by stages we have:

(a) $2P \times 5 = 10P$
(b) $10P + 25$
(c) $(10P + 25 + L)10 = 100P + 250 + 10L$
(d) $100P + 250 + 10L + W$
(e) $100P + 10L + W$

so that the number 'PLW' must result.

EXAMPLE 6

For this example, a person is asked to choose a number of five or more digits. He is then to add the digits and to deduct this sum from the original number. In the number which results he is to cross out one of the digits other than a nought and, having done this, to add the remaining digits together and to disclose the result. If this result consists of more than one digit the procedure for the extraction of digital roots (as explained in Chapter 4) must be followed until the final root has only one digit. This digital root is then deducted from 9 and the balance will be equivalent to the digit crossed out. The performer is therefore able to 'divine' the digit which was crossed out even though he does not know the original number selected.

If, for instance, the number originally selected is 12345, the sum of the digits is 15. If this digital sum is deducted from the original we have $12345 - 15 = 12330$; and if the digit 3 is crossed out, the remaining digits total 6. This total, deducted from 9, gives the digit which was crossed out, 3.

The reason for this is that if we deduct from any number the sum of its digits the result is always a multiple of 9, so that the digital root of the result is also 9.

The elimination of any digit in the result therefore reduces the digital root by an equivalent amount to the digit crossed out.

The effect of this manœuvre may be made more mystifying if the person who selects the number is not also asked to add up the remaining digits. If he is told to read out these digits, instead of their sum, the performer may add them mentally and proceed as before.

The foregoing examples are all similar in that they do not require any equipment for their performance. Selected numbers may also be discovered by the use of specially prepared cards, each card bearing certain sets of numbers. In the example shown on the next page there are six sets of numbers.

Any number from 1 to 63 may be found by adding together the first numbers in each set in which the particular number appears. The number 63 appears in each set and the total of all the first numbers in each set is

$$1+2+4+8+16+32=63$$

Anyone who selects a number is therefore asked to state in which sets of numbers it appears. The addition of the first numbers in the sets mentioned will give the number selected.

Set A			Set B			Set C		
8	27	46	1	23	45	16	27	54
9	28	47	3	25	47	17	28	55
10	29	56	5	27	49	18	29	56
11	30	57	7	29	51	19	30	57
12	31	58	9	31	53	20	31	58
13	40	59	11	33	55	21	48	59
14	41	60	13	35	57	22	49	60
15	42	61	15	37	59	23	50	61
24	43	62	17	39	61	24	51	62
25	44	63	19	41	63	25	52	63
26	45		21	43		26	53	

Set D			Set E			Set F		
4	23	46	32	43	54	2	23	46
5	28	47	33	44	55	3	26	47
6	29	52	34	45	56	6	27	50
7	30	53	35	46	57	7	30	51
12	31	54	36	47	58	10	31	54
13	36	55	37	48	59	11	34	55
14	37	60	38	49	60	14	35	58
15	38	61	39	50	61	15	38	59
20	39	62	40	51	62	18	39	62
21	44	63	41	52	63	19	42	63
22	45		42	53		22	43	

The 'Counting' Cards. Any number may be made up by adding together the first numbers on each card, or set, on which the number appears. Thus: $47=8+1+4+32+2$.

Another basic number property which may be made the basis for similar amusement is as follows. Take any number of two digits and reverse the digits. Then the difference between the two numbers thus formed will be nine times as great as the difference between the two digits used. The difference between 92 and 29 is 63, and this is nine times greater than 7 which is the difference between the two digits 2 and 9.

As is usual in these relationships, the proof is elementary. The original number may be expressed in the form $10x+y$, where x and y are the digits used, and the second number is $10y+x$. Their difference is therefore

$$(10x+y)-(10y+x)=9(x-y)$$

From the foregoing examples it should now be quite clear that when a person appears to discover the nature of a number by allegedly magical processes, he is in reality mentally working out a mathematical relationship and merely supplying his audience with the result.

On the other hand, the mental ability of some individuals is itself an apparently magical quality. From time to time there have appeared certain persons who, because of their ability to perform remarkable feats of mental arithmetic, earned the title of 'calculating prodigies'. These individuals, without any undue preparation, have proved themselves equal to various mathematical problems which they solved by a mental process in a matter of seconds, whereas even the more accomplished mathematicians found their powers fully extended to approach anywhere near the same speed with the aid of written calculation.

Few of these prodigies have ever been able to give a satisfactory explanation of the methods of calculation used, yet many of their achievements can be described only as phenomenal. There are records of one person who could mentally extract the square root of a number consisting of 100 digits and, as if that feat alone were not sufficiently remarkable, the very act of memorizing a hundred digits in their correct order while he subjected the number to whatever process he used would itself be memorable.

An excellent memory is, of course, essential to any process

of mental calculation and the example just quoted is typical of many. At a public performance a boy of ten years of age was given a number which was read backwards to him. He immediately repeated it in its proper order and was able to repeat it correctly an hour later. The number contained 43 digits. The reader is invited to test his own memory in the same way, and then to remember that, whatever the degree of success he achieves, the feat of memorizing is but the first step and that the true prodigy is able to memorize numbers instantly.

It is a consoling thought for the vast majority of us that these wonder men of numbers have been extremely rare and of a race apart, and that despite their remarkable powers they have often proved to be uneducated in other arts. Some of them, indeed, could neither read nor write.

14

Fallacies

Mathematics is an exact science. The number 2, multiplied by the number 3, will always give a result equivalent to the number 6. That is incontrovertible fact, but there are numerous pitfalls for the unwary. In all scientific laws there is the proviso: 'all other things being equal'. Numbers will always behave towards each other in a fixed and certain manner provided they are treated purely as numbers; when they are used to represent something else, this 'something' may introduce certain complicating extraneous factors.

In mathematics, as all who have struggled through examinations will agree, everything is not always what it seems. Thus, although it is a fact that twice times x will always be equivalent to $2x$, it does not necessarily follow that the doubling of an excise rate on tobacco will inevitably result in the doubling of the total revenue derived from the tax. This is because the amount of revenue will depend upon the *amount* of tobacco purchased as well as on the actual excise rate. The raising of the *rate* of tax might conceivably reduce the total revenue if it resulted in a heavy drop in tobacco consumption.

In the same way, it must be remembered that the use of mathematical devices for the solution of problems will give ridiculous results, when applied to practical matters, if certain factors are ignored. If a man can plough two fields in one day, then two men—with equal facilities and at equal speeds—will be able to plough the same two fields in half a day. In the circumstances quoted, the doubling of the labour force will halve the time taken to complete the task in hand. It does not necessarily follow however that twenty men will be able to plough the same two fields in one-twentieth of the time, for the simple reason that there may not be room in the two fields for so many ploughs. Even where the toil is of a manual nature the argument is the same. Three men will be able to make a concrete path in three hours, but ninety men would not, as a consequence, be able to make the same path in $\frac{9}{90}$ths of an hour (i.e. 6 minutes). Even allowing for the shovel-leaners, tea-makers and card-players of this large labour force, the remainder of the path-makers would still be so numerous that they would constantly be falling over each other.

In Algebra and other branches of Mathematics, we use certain letters and other symbols to replace unknown quantities and relationships. If these can be placed into the form of an equation containing only one unknown quantity, then that quantity may be evaluated simply by working out the equation. Similarly, the values of two unknown quantities may be calculated from simultaneous equations. But the use of letters is intended to simplify these calculations; if we are not careful, they serve only to confuse. Indeed, certain equations may be expressed in an algebraic form which appears to be correct but which, in fact, may contain hidden fallacies leading to fantastic results.

The best known examples of such results are those involved in 'proving' that a number greater than nought is nevertheless equal to nought or to any other desired number. This spoofing is a simple matter and may be accomplished in a number of different ways.

The following is one demonstration, in the case where two unknowns x and y are found to be equal to each other—and, therefore, their squares are also equal.

Thus: $x = y$
Therefore $x^2 = xy = y^2$
and $x^2 - y^2 = x^2 - xy$
Factorizing: $(x-y)(x+y) = x(x-y)$
so $x + y = x$

FALLACIES

But, as $y=x$, then $\quad x+y=x+x$
so $\quad\quad\quad\quad\quad\quad 2x=x,\ \text{and}\ \ 2=1$
Similarly $\quad\quad x^2+xy-2y^2=x^2-xy$
Factorizing: $\quad (x-y)(x+2y)=x(x-y)$
therefore $\quad\quad\quad (x+2y)=x$ [Dividing by $(x-y)$]
But, again, $y=x$, so $\quad 3x=x$
and $\quad\quad\quad\quad\quad\quad 3=1$

The fallacy is not difficult to find. In each of the above examples each equation, after factorization, was divided by $(x-y)$. But if $x=y$, then $x-y=0$. When we divide, say, 2 by nought, the answer, so far as can be stated briefly, is infinity; the answer is certainly not 2. Similarly, when we divide $x(x-y)$ by $(x-y)$ the answer cannot be x when $(x-y)=0$. In point of fact, therefore, when we divide by $(x-y)$ we are claiming to do the impossible.

There are also a number of equations for x which hold when $x=1$, but not when x is greater than 1.

Example: $\quad\quad\quad\quad x^4+x^3=2x$

If x is equal to 1, then this is obviously correct, but if $x=2$, then it is not. This, of course, is because the number 1 multiplied by itself any number of times never varies; it always remains 1. This is the only number which acts in this way. how about zero

The use of the minus sign can also result in fallacious results, as here:

$$(-5)\times(-5)=25$$
and $\quad (+5)\times(+5)=25$
so $\quad\quad (-5)\times(-5)=(+5)\times(+5)$
or $\quad\quad\quad (-5)^2=(+5)^2$

and, taking the square root:

$$-5=+5$$
so that $\quad -5-5=0$
or $\quad\quad\quad 0=+5+5$

The fallacy here is in taking the square root of $(-5)^2$ and $(+5)^2$. Both these expressions are really equivalent to 25, but the square root of 25 is neither $+5$ nor -5 alone. The correct root is *plus or minus 5*.

The minus sign is also featured in the following. Let x be the quantity which satisfies the equation $a^x = -1$. By squaring both sides we obtain $a^{2x} = 1$. But a^0 always equals 1, no matter what value a represents. Thus, $a^x = a^0$, so that $2x = 0$, and $x = 0$.

But, again $$a^0 = 1$$

and, if x equals nought, then a^0 may also be shown as a^x.

Therefore $$a^x = +1$$

But, from our data: $$a^x = -1$$

so $$+1 = -1$$

It has been noted that $a^0 = 1$, irrespective of the value of a, and that, therefore, $$1^0 = 2^0 = 3^0$$

but it does not follow from this that:

$$1 = 2 = 3$$

Similarly $\dfrac{a-b}{a-b} = 1$; except where $a = b$, for then $(a-b) = 0$, so that $\dfrac{a-b}{a-b}$ has no meaning in fact.

In the same way, the expression $y = \dfrac{0}{x}$ appears to be meaningless since it is impossible to divide nought by x. The expression is not, however, entirely devoid of meaning. It may be re-expressed as:

$$xy = 0$$

whence we know that *either x or y* must be equal to nought. This fact is a fundamental one and the solution of quadratic equations depends largely upon it. Thus, where

$$(x-1)(x+1) = 0$$

we can state that either $(x-1)$ or $(x+1)$ is equal to nought and that, therefore:

$$x = +1 \quad \text{or} \quad -1$$

The next example of fallacious reasoning involves the understanding of logarithms.

$$\log(1+x) = x - \tfrac{1}{2}x^2 + \tfrac{1}{3}x^3 - \tfrac{1}{4}x^4 + \tfrac{1}{5}x^5 - \ldots \text{ etc.}$$

[The logarithms are to the base e.]

If $x=1$, then $\log(1+x)$ becomes $\log 2$ and

$$\log 2 = 1 - \tfrac{1}{2} + \tfrac{1}{3} - \tfrac{1}{4} + \tfrac{1}{5} - \tfrac{1}{6} + \tfrac{1}{7} - \tfrac{1}{8} + \text{ etc.}$$

Multiply this by 2:

$$2 \log 2 = 2 - 1 + \tfrac{2}{3} - \tfrac{1}{2} + \tfrac{2}{5} - \tfrac{1}{3} + \tfrac{2}{7} - \tfrac{1}{4} + \text{ etc.}$$

Collect all terms with common denominator in order of size:

$$\begin{aligned}2 \log 2 &= (2-1) - (\tfrac{1}{2}) + (\tfrac{2}{3} - \tfrac{1}{3}) - (\tfrac{1}{4}) \quad \text{etc.}\\ &= 1 \quad\quad - \tfrac{1}{2} + \quad \tfrac{1}{3} \quad - \tfrac{1}{4} \quad \text{etc.}\end{aligned}$$

which is the same as $\log 2$.

Therefore $\quad\quad\quad\quad \log 2 = 2 \log 2$

and again $\quad\quad\quad\quad\quad 1 = 2$

This fallacy is due to the rearrangement of the terms, not all of the same sign, of an infinite series.

Infinite series also afford another fallacious example.

Let $\quad \tfrac{1}{1} + \tfrac{1}{3} + \tfrac{1}{5} + \tfrac{1}{7} + \tfrac{1}{9} + \tfrac{1}{11} + \ldots = x$

and $\quad \tfrac{1}{2} + \tfrac{1}{4} + \tfrac{1}{6} + \tfrac{1}{8} + \tfrac{1}{10} + \tfrac{1}{12} + \ldots = y$

then $\quad \tfrac{1}{1} + \tfrac{1}{2} + \tfrac{1}{3} + \tfrac{1}{4} + \tfrac{1}{5} + \tfrac{1}{6} + \ldots = 2y$

but $\quad\; \tfrac{1}{1} \;+\; \tfrac{1}{3} \;+\; \tfrac{1}{5} + \ldots \quad\quad = x$

therefore, subtracting:

$$\tfrac{1}{2} + \tfrac{1}{4} + \tfrac{1}{6} \quad\quad = 2y - x$$

but $\quad\quad \tfrac{1}{2} + \tfrac{1}{4} + \tfrac{1}{6} \quad\quad = y$

therefore $\quad\quad\quad 2y - x = y$

and $\quad\quad\quad\quad\quad y = x$

so that $\quad\quad\quad\quad x - y = 0$

Now the expression $x - y$ is equivalent to:

$$\tfrac{1}{1} + \tfrac{1}{3} + \tfrac{1}{5} + \tfrac{1}{7} + \tfrac{1}{9} + \tfrac{1}{11} + \ldots - \tfrac{1}{2} - \tfrac{1}{4} - \tfrac{1}{6} - \tfrac{1}{8} - \tfrac{1}{10} - \tfrac{1}{12} -$$

which, by regrouping, may be converted to:
$$\tfrac{1}{1}-\tfrac{1}{2}+\tfrac{1}{3}-\tfrac{1}{4}+\tfrac{1}{5}-\tfrac{1}{6}+\tfrac{1}{7}-\tfrac{1}{8}+\tfrac{1}{9}-\tfrac{1}{10}+\tfrac{1}{11}-\tfrac{1}{12}+ \ldots$$
and is in fact approximately equivalent to ·7, being a non-terminating expression with a limiting value.

The fallacy here is that both x and y are infinite and infinity cannot be subjected to normal mathematical processes. You cannot mathematically add infinity to infinity for the result would still be—infinity.

In Chapter 11, it was shown that certain successive operations upon a three-digit number would always give the result of 1089. A proof that this must be so was included in the same chapter. Any attempt to prove it, however, without bearing in mind that a is greater than c and that it is necessary to employ the strategy of borrowing, would give the following quite fallacious result.

Original number	$100a$	$+10b+$	c
Reverse	$100c$	$+10b+$	a
Subtract	$100(a-c)$	$+\ 0\ +$	$(c-a)$
Reverse	$100(c-a)$	$+\ 0\ +$	$(a-c)$
Add	$100(a-c+c-a)+$	$0\ +$	$(c-a+a-c)$
	$=100(0)$	$+\ 0\ +$	(0)
	$=0$		

Care is also necessary when dealing with comparative ages. Thus when Brown is 30 and Smith is 10, Brown is three times as old as Smith. But in ten years time, when Brown is 40 and Smith is 20, Brown will only be twice as old as Smith. No time need be wasted in calculating how long Smith and Brown must live before they both become the same age!

A very old problem was posed thus: An old man in his will left 17 horses to be divided amongst his three sons in such a way that they should receive one-half, one-third and one-ninth of the total respectively. As the number 17 is not divisible by 2, 3 or 9, how was the division (in terms of whole horses) to be effected? The solution to this problem was provided by the ingenious notion of borrowing another horse (on paper only). This made the total of horses eighteen instead of seventeen, and of this new total:

$$\frac{1}{2} = 9$$
$$\frac{1}{3} = 6$$
$$\frac{1}{9} = 2$$

These are the respective shares for the three sons, and these shares total 17 (i.e. 9+6+2), leaving the borrowed horse to be 'paid back'.

It is to be noted that each son receives a fraction of a horse more than his entitlement, because the father bequeathed less than his full possessions. This explains the apparently fallacious argument that all three sons may receive their full legacies and yet leave the borrowed horse to be returned to its owner. The bequests were $\frac{1}{2} + \frac{1}{3} + \frac{1}{9}$ and these total only $\frac{51}{54}$ths of the 17 horses. The extra 'parts' received by each son are in proportion to his legacy. A has half a horse extra; B has an extra third and C has an extra ninth of a horse.

This problem is not confined to the numbers used above, but may be posed and solved by the same method where other numbers are involved. If, for example, there were 19 horses to be divided amongst A, B and C in the proportions, one-half, one-quarter and one-fifth, we can, by adding 1 to the total, divide the new total into the required amounts and still have the extra horse left over afterwards. The new total would be 20, and these would be distributed as follows:

$$\frac{1}{2} = 10$$
$$\frac{1}{4} = 5$$
$$\frac{1}{5} = 4$$
$$\text{Remainder} = 1$$

There are many similar examples.

Fallacies usually arise in numbering because of misinterpretations of the meanings of certain algebraic or purely numerical identities, but nowhere is there likely to be greater misinterpretation than in the construing of statistics.

In modern life, statistics are at the very heart of most of our institutions. Everything is reduced to an average. The cost of living index is based upon an average person's probable expenditure; polls of opinion are recorded from a selected representative cross-section of the public; and cricketers, to mention just one example, have tables compiled showing their average number of runs per innings.

In these, and in many other ways, numbers are accumulated and treated (and often maltreated) in a number of different ways. Their value lies primarily in showing trends or in providing comparisons rather than in demonstrating anything of a fixed nature, and also depends largely upon the representative nature of the bases upon which they are calculated. For example, a cricketer's average may be shown as 34·5 runs per innings even though it is quite impossible to score ·5 of a run in any circumstances. This average may, however, be compared with his averages at other times and also with the averages of other players, and so provides useful comparisons.

All statistics have their inherent faults and must be treated severely on their merits. A consensus opinion poll, carried out among a thousand individuals, is no reliable guide to the opinion of a nation, no matter how apparently representative is the selection of the individuals questioned. And again, the average person, upon whose estimated expenditure so many statistics depend, just does not exist. Nevertheless these details do, when considered in the proper manner, give clues to trends in income and expenditure which are of immense value to the economist.

Once the statistics are accumulated, emphasis then rests upon the correct presentation and understanding of them. The cynic will tell you that figures will prove anything, and it is true that although there may be only one set of figures, they can be so presented in different ways as to purport to show quite different trends, in the same way as photographs of the same room, taken from different angles, appear to reproduce different rooms.

FALLACIES 143

There are many examples of this dual nature in statistics whenever politics come into the question. For instance, under the present British electoral system, it is possible for Party A to hold more seats in Parliament than does Party B and yet, because of the system of electing members, to have a smaller aggregate of actual votes. It is obvious that if Party A are preparing statistics they will omit all reference to the aggregate number of votes, whereas Party B would lay heavy stress upon that figure.

Another set of numbers which require close attention are those expressed as percentages of others. It may sound childish to say that a number will be equivalent to different percentages of different numbers, yet it is a common error for these to become confused. For example, the profit obtained from the sale of an object is a certain percentage of its cost price but a totally different percentage of its selling price, because the selling price also includes the profit. To express the profit as a percentage of the selling price therefore gives quite a false impression of the true facts.

Other misconceptions arise in calculating cost or selling price if one of these prices and the rate of profit are known. If the cost price is £100 and the profit is 25 per cent. on cost price, then the selling price is £100+£25=£125. That is, to obtain the selling price we add a quarter of the cost price. If, however, we start with the selling price, we cannot deduct a quarter of the latter in order to arrive at the cost price. We have to bear in mind that the selling price of £125 is equivalent to cost price plus profit; that is, it is one-and-a-quarter times the cost price, and we have to deduct only one-fifth:

$$£125-£25=£100$$

Percentages are also very misleading in connexion with the declaration of companies' dividends in the allocation of trading profits. The unwary investor who hears that certain shares are yielding 50 per cent. dividends may think that if he invests £100 in those shares, then in two years he should be able to double his capital. What has to be remembered, however, is the fact that dividends are declared on the nominal value of the shares.

Thus, if the nominal value of the share is £1 and a dividend of 50 per cent. is declared, then the dividend in respect of that share is ten shillings. The unfortunate investor will probably find, however, that he cannot buy any of the necessary shares at their nominal value. In fact he may have to pay £5 for a share. If he does so, he will still receive only the dividend of ten shillings, a return of 10 per cent. on his actual outlay. This is a very different proposition, and yet limited fortunes may still be made on the Stock Exchange!

15

Magic Squares

A Magic Square is composed of a number of integers so arranged within the square formation that the sum of the integers in each row, each column and each diagonal is identical. To deal comprehensively with Magic Squares would require almost a volume by itself, and it is necessary to confine consideration here to the simpler principles involved.

Two examples are given below, showing different ways in which the consecutive numbers from 1 to 16 may be arranged so as to form magic squares.

16	3	2	13
5	10	11	8
9	6	7	12
4	15	14	1

15	10	3	6
4	5	16	9
14	11	2	7
1	8	13	12

In both squares, each row, column and diagonal gives a total of 34. It will also be observed that:

(a) if each square is subdivided into four smaller squares, each containing four numbers, the total of each smaller square is also 34;
(b) in the second square, further smaller squares, each

totalling 34, may be made by taking any four numbers in square formation to each other, thus:

4	5
14	11

11	2
8	13

This property is not found in the first square;

(c) the four central numbers in each square also total 34.

In every magic square the total of every row, etc., is directly related to the number of integers used. It is quite apparent, of course, that if the lowest integer used is 1 and all the integers used are consecutive, then the highest integer (and thus also the total number of integers used) must be a square number of the form n^2.

The total obtained by adding together all the integers used is of the form:

$$\frac{n^2(n^2+1)}{2}$$

being the sum of an arithmetical progression whose first term is 1, whose last term is n^2 and whose common difference is 1. This total has to be spread equally over n rows (or, looked at from another angle, over n columns) and the total of each row or column is therefore:

$$\frac{n^2(n^2+1)}{2n} = \frac{n(n^2+1)}{2}$$

In the two example squares shown, there are 16 consecutive integers used. For this purpose, therefore, $n^2 = 16$, and $n = 4$. The total of each row, column or diagonal is therefore required to be $\frac{4 \times 17}{2} = 34$, as shown.

There are a number of rules which enable one to construct a magic square without much difficulty, and the rules vary according to whether the number of integers used is odd or even.

A square, not greatly different from the first one shown above, may be built up as follows. First, it is necessary to construct the framework of square, sub-squares and diagonals, as in Fig. (i).

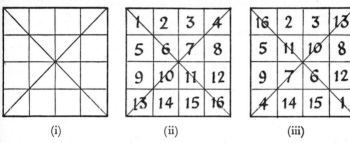

(i)   (ii)   (iii)

Next, write in the integers in their normal order from left to right in horizontal rows, as in Fig. (ii). In the final stage (Fig. iii), integers in sub-squares which are crossed by diagonals change places with their diametrically opposite integers on the same diagonals (e.g. the integers 13 and 4 change places), but the integers in the other sub-squares remain unaltered. Fig. (iii) is then a magic square.

If the number of integers in a magic square is to be odd, the square may be constructed in the following manner:

(a) Write the first number in the middle sub-square of the first row.

(b) Write the second number in the following column, but in the bottom row, thus:

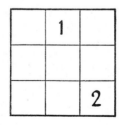

(c) Continue writing numbers consecutively in a diagonal line sloping upwards from the previous number and to the right-hand side of the square. When the last column on the right-hand side is reached, the next highest number is to be written in the extreme left-hand column as if it followed immediately after the right-hand column, and the process is continued, thus:

		1		
	5			
4				
				3
			2	

(d) When a sub-square next in line is reached but has already been filled, the 'path' of the numbers drops to the sub-square next below and then begins to climb diagonally again from that point. When the top row is reached, the next number is written in the bottom row as if it followed immediately above the top row. Following these rules gives the following complete square where 25 integers are used.

17	24	1	8	15
23	5	7	14	16
4	6	13	20	22
10	12	19	21	3
11	18	25	2	9

MAGIC SQUARES

A very different construction for a square of the same size, or order, as the one above is by means of two preliminary squares. In square A, the numbers 1 to 5 inclusive are placed in the top row in any order preferred. The second row begins with the number in the fourth sub-square of the first row and this is followed by the remaining numbers in the same relative order. Each successive row commences with the number in the fourth sub-square of the previous row.

3	5	4	1	2
1	2	3	5	4
5	4	1	2	3
2	3	5	4	1
4	1	2	3	5

Square A

Square B is built up from the numbers 0, 5, 10, 15 and 20. These numbers are placed in the top row in any order. Each successive row starts with the number in the third sub-square of the previous row, and the other numbers then follow in the same relative order.

0	10	20	5	15
20	5	15	0	10
15	0	10	20	5
10	20	5	15	0
5	15	0	10	20

Square B

The numbers in the same relative positions in squares A and B are now added together, giving square C, which is the final square including all the integers from 1 to 25.

3	15	24	6	17
21	7	18	5	14
20	4	11	22	8
12	23	10	19	1
9	16	2	13	25

Square C

It will be noticed that all three squares (A, B and C) are magic.

The following square differs from others in that it contains a smaller magic square formed by the nine central digits.

2	23	25	7	8
4	16	9	14	22
21	11	13	15	5
20	12	17	10	6
18	3	1	19	24

Another method of using a preliminary square to obtain a final square which shall be magic, is to construct a larger square balanced on one of its angles, with the framework of the required square drawn inside the first square so that the sides of the first square are bisected by the corners of the smaller square. Numbers are then written consecutively in the first square, as follows:

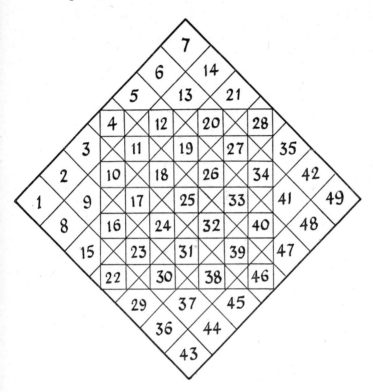

It will be seen that some of the integers automatically appear in the smaller square, because the sub-square of the two main squares partly coincide in places. The numbers which appear in the smaller square are left in their respective positions, but the numbers which appear outside the smaller square are transferred to positions within that square. Each such number is, in fact, moved to the furthest vacant sub-square along the same row or column of the larger square.

The number 3, for instance, is transferred to the last sub-square in the second row.

The following magic square results:

4	29	12	37	20	45	28
35	11	36	19	44	27	3
10	42	18	43	26	2	34
41	17	49	25	1	33	9
16	48	24	7	32	8	40
47	23	6	31	14	39	15
22	5	30	13	38	21	46

Magic squares are by no means confined to the use of consecutive integers. They may, in fact, be composed from any set of integers which conform to certain rules. A square may be made from nine integers (*a*) if these integers can be written in three sets of integers, each set being an arithmetical progression with a common difference of 1, and (*b*) if the relative terms in these sets can be extracted to form new progressions with a common difference of 6. The following nine integers may be so arranged:

1 2 3 7 8 9 13 14 15

and their arrangement is:

	A	B	C
(i)	1	2	3
(ii)	7	8	9
(iii)	13	14	15

Each of the horizontal sets, (i), (ii) and (iii), form progressions with a common difference of 1. Each of the vertical sets, A, B and C form progressions with a common difference of 6.

These numbers may be formed into a magic square by means of the procedure outlined for the construction of squares of an odd order. The following square results:

14	1	9
3	8	13
7	15	2

Another way of forming a magic square is now shown. For a square of the third order (i.e. where $n=3$), first construct two squares, and letter the sub-squares of one as under. This latter square is then the key square and the first square is built into a magic square by reference to the letters in the key square.

Key Square

A	B	C
D	E	F
G	H	K

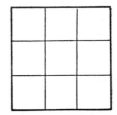

Next, we select any three numbers. Emphasis is laid upon the fact that it does not matter which numbers are chosen. For our example, we shall use the numbers 5, 9 and 47. One of these numbers (say, 47) is placed in the sub-square relative to sub-square B in the key square. Add this same number to either of the other two selected numbers (say, 5) and place the total in sub-square G. Add the second number again to the number just placed in G and place the new total in F. This gives us:

	47	
		57
52		

in which the square contains three numbers. We now treat each of these numbers in turn. Add the third number (9) to the first number (47) and place the total in sub-square K. Add 9 again and write the result in D. Add 9 to the number in G and place the result in E. Again add 9 to the number written in E and write the result in C. Now add 9 to the number in F and place the result in A. Add 9 again and place the result in H. All sub-squares are now filled and we have:

66	47	70
65	61	57
52	75	56

The reason why this must be a magic square is best demonstrated by showing how each individual number is obtained.

Other squares have been constructed using only prime numbers, and it is possible to construct what are called doubly-magic squares, the peculiar property of which is that, if the original integers are squared, the resulting square is also magic.

Finally, we may refer to subtracting and multiplying magic squares. In the former type of square, a constant 'total' results for each row, column, or diagonal by deducting

MAGIC SQUARES

47+5+ 5+9	47	47+5 +9+9
47+ 9+9	47+ 5+9	47+ 5+5
47 + 5	47+5+ 5+9+9	47 + 9

its first digit from the second and the result from the third digit and so on. An example of such a square is:

2	1	4
3	5	7
6	9	8

In this square, the constant is 5. It should be noted that, in practice, it is simpler to deduct the middle integer in each row from the other two integers in the same row. The same result is obtained.

In a multiplying square, all the integers in each row, column or diagonal, when multiplied by each other, give a constant value, as under:

12	1	18
9	6	4
2	36	3

In this instance, the constant is 216.

16
Number Mystery

Because of the regular and changeless relationships observable between the integral numbers, the ancient philosophers claimed to be able to discover a mystical significance in many of them—particularly the nine digits—individually. Mathematical truths were accepted as fixed and certain numbers, being representations of these truths, were regarded also as being revelations from the controlling deities.

The number *one* was generally considered to have a particularly sacred identity. Because of its indisputable position as the origin of all numbers it was held to be the numerological symbol of life and creation, and it derived added strength and stature by reason of its indivisibility.

The number *two* was accounted the first feminine number; all even numbers being considered feminine and all odd numbers masculine. In general the odd numbers were thought to exercise favourable influences whereas the even numbers were, somewhat unchivalrously, considered unfavourable. Consequently the number *two* found very little favour and, indeed, was often regarded as a symbol of death and evil generally.

In the Pythagorean philosophy, the number *three* was considered the acme of perfection because it expressed the beginning, the centre and the end of all things. It was the universe in a nutshell! This number was extremely popular when numerology was a flourishing philosophy and its popularity has lasted to the present day. It appears repeatedly in both the Old and New Testaments of the Bible, its most significant function being the symbolical representation of the Holy Trinity.

Another biblical association is provided by the three wise men bearing their gifts, whilst in classical mythology there were three Graces and three Furies. In heraldry we find the three British lions in the royal arms; three legs in the quaint device of the Isle of Man; and the three feathers of the Prince of Wales' Feathers. In modern times we are all familiar with, if not proficient in the use of, the three R's of basic education; we are often called upon to give three cheers in some cause or other and, in our nursery days, most have lived with the three blind mice, the three bears and the three little kittens who lost their mittens.

Four was also a sacred number to the Pythagoreans because it was the first square number. It was deemed to represent the four elements and was, as a consequence, the symbol for the Earth.

The number *five* was the number of knowledge. Its symbol was the pentagram and this was also often taken as the symbol of Man, representing his five members (head and four limbs) and his five fingers on each hand. *Five* was also the symbol of marriage because it expressed the union of the odd number, three, and the even number, two.

Equilibrium and peace were represented by the number *six*, its symbol being two triangles base to base. This identity is still expressed in the phrase 'sixes and sevens' to describe the upsetting of the equilibrium, and also in the phrase 'six to one and half a dozen to the other' to emphasize a balancing equality.

The number *seven* rivalled the number three in popularity. It has repeated references in the Bible, having particular significance in connexion with the day of rest following the creation of the World. Then there are also the Seven Virtues, the Seven Deadly Sins and the Seven Sleepers. There were seven wonders of the ancient world and, throughout the years, men have yearned to attain the ecstasies of their seventh heaven.

To the Pythagoreans, the number *eight* was the number of justice because it divided evenly into four and four, and each of these divided evenly into two and two, which in turn, divided evenly into one and one. The same properties exist of course in any power of the number two, so that it is difficult to understand why the number eight should have been especially selected. But, as being the first cube number, it was also regarded as the corner-stone and capacity and was therefore the symbol of plenty.

The number *nine*, being the result of multiplying the number three, a favourable number, by itself, was also believed to be favourable. It was represented by three triangles and thus was taken to symbolize the equilibrium of the three worlds. It is not known what connexion, if any, such symbolism demonstrates between the nine lives of a cat, a nine-days' wonder or the nine points of the law.

The number *ten* was held in reverence as the symbol of the Absolute and of Law, while *eleven* was deemed to be evil, being a transgression of the number ten. The number *thirteen* has always been thought to be unlucky and is still strongly believed by many people to exercise a baneful influence over their fate—a belief which is exemplified by the fact that, in some hotels, the room which would normally be numbered 13 is instead numbered as 12A.

Because of their own special properties, the first two Perfect numbers, *six* and *twenty-eight*, were accorded particular importance. Their significance is exemplified respectively by the six days of creation and the twenty-eight days in each lunar month.

The influence of mysticism associated with numbers still survives in folklore. The seventh son of a seventh son is supposed to possess remarkable healing powers. The common cold is supposed to run a course of nine days—three days coming, three days at its peak, and three days going. The human body is said to change completely every seven years and, even in law, an infant becomes an adult at the age of twenty-one (that is, three times seven).

In the Bible, the Apocalyptic number 666 is given as the number of the Beast, generally interpreted as the Antichrist.

NUMBER MYSTERY

Various people at different times and by varying numerological processes have claimed to be able to prove that this number represented their particular enemies. It may be mentioned that the only apparent fact known about the number is that it is triangular and palindromic. It is also the sum of all consecutive integers from 1 to 6^2.

There are at least two methods whereby modern numerologists seek to prove that every person has an affinity for certain numbers or, vice versa, that certain numbers 'vibrate' favourably or unfavourably with regard to particular individuals.

One method is to reduce the individual's date of birth to its digital root. Thus the birth-date 12.7.1932 would reduce to the digital root of 7, unless the date is shown alternatively as 12.7.32 (as in some systems) when the root would be 6. The digital root is then termed the significant number in respect of that individual.

From this significant number, the numerologist claims to have the power to divine the nature and character of the individual, his fortunes in the future and, amongst other intimate details, the significant number of the lady he should seek in marriage!

The second method is to assign a numerical value to every letter in the alphabet and then to express the individual's name in numerical form, again reducing the latter to its digital root. The interpretation of the significance of this root then proceeds as before.

Superficially it would appear that the numerologists claim to discover a great deal of significance within the limited scope of nine digits. Since the decimal system of counting is of only an arbitrary nature, the reader may be excused for wondering what developments in numerology would result from the adoption of a new scale of notation.

If, for instance, the scale of *twelve* were to be adopted and two new digits were to emerge for the numbers 10 and 11 respectively, would there be an automatic increase in the amount of numerologically significant numbers and, as a result, a similar increase in the number of different divisions of humanity? For if there were no such increases, it must be presumed that those individuals whose significant numbers reduced to the roots of 10 or 11 would have no future at all!

It has even been seriously suggested that a person might be able to change his fortunes merely by changing his name, or by using initials instead of his full name, and thereby altering his significant number. Modern numerology would appear to involve a process of thought which compensates by its adaptability for what it lacks in logic.

The dangers inherent in any system which ascribes a mystic significance to numbers may perhaps be best exemplified in a modern hypothetical case—that of the number 999 which is dialled on the English telephone services for emergency calls. This number has taken to itself a significance quite divorced from its origin as a telephone number, and has now come to be strictly identified with the police and legal processes, despite the fact that it is also used for emergency ambulance and fire service calls as well.

Perhaps we have here the seeds of a future belief that the number 999 is a numerological symbol for the maintenance of law and order against the powers of evil. Arguments of the type normally forthcoming could easily be adduced to support such a belief. The number 999 is odd and therefore masculine and favourable; it consists entirely of the repetition of the digit 9, a favourable number, and its digital root also reduces to 9; and, still more significant, it is the number of the Beast upside-down!

All this in the centuries to come may be 'proved' about a number, which really owes its significance to the very material fact that it consists of the repetition of the one digit which occupies a particular position on a telephone dial and may therefore be dialled in the dark, being located by the sense of touch alone.

As a final thought we may consider that *nine* is not only one less than *ten*; it is also equal to one over the eight and, seen in this light, it is apt to lose much of its respectability.

APPENDIX I

Pascal's Triangle and the Binomial Theorem

The following triangle shows at a glance the coefficients of the expansion of any binomial of the form $(x+1)^n$.

```
                    1    1
                 1    2    1
              1    3    3    1
           1    4    6    4    1
        1    5   10   10    5    1
     1    6   15   20   15    6    1
  1    7   21   35   35   21    7    1
```

These coefficients are derived from the general formula for the expansion of

$$(x+a)^n = x^n + nx^{n-1}a + \frac{n(n-1)}{1 \cdot 2} x^{n-2}a^2 + \ldots a^n$$

If $a=1$, then all powers of a are also equal to 1. Any number multiplied by 1 remains unchanged, so that a and its powers may be eliminated from the expansion. Then

$$(x+1)^n = x^n + nx^{n-1} + \frac{n(n-1)}{1 \cdot 2} x^{n-2} + \ldots + 1$$

and the coefficients of this expansion are, in order:

$$1 \quad n \quad \frac{n(n-1)}{1 \cdot 2} \quad \text{etc.} \ldots \quad 1$$

it being remembered that n represents the value of the power to which the binomial is being raised. Thus, for different powers, the coefficients are:

Power	Coefficients
1	1 1
2	1 2 1
3	1 3 3 1
4	1 4 6 4 1

and the construction of the triangle becomes apparent.

Once the building of the triangle has been started, however, there is no need to use the full formula in order to ascertain the coefficients for any specified expansion. The required coefficients may be derived from the triangle itself. The first and last coefficients are both invariably 1. The other coefficients of any one expansion may be derived from the expansion one degree lower—merely by adding together the coefficients of the latter taken in pairs. The coefficients of the expansion where $n=4$, are:

$$1 \quad 4 \quad 6 \quad 4 \quad 1$$

Taking these in pairs, we have:

$$1+4 = 5 \quad 4+6 = 10$$
$$6+4 = 10 \quad 4+1 = 5$$

and the coefficients of the expansion where $n=5$ are therefore:

$$1 \quad 5 \quad 10 \quad 10 \quad 5 \quad 1$$

Another binomial expansion may be employed in a method for the extraction of the roots of numbers. In the expansion of $(1+x)^n$ where n is integral and positive, all the terms of the expansion are also positive, but when n is fractional and less than unity, the terms of the expansion which contain powers of x are alternately positive and negative. This distinction need not concern us here, however. The expression $(1+x)^n$ may be calculated *approximately* by taking only the first two terms of its expansion, if x is very small. In these circumstances, the subsequent terms are too small to have any significant effect on the result.

Thus: $(1+x)^n = 1+nx$ approx.

PASCAL'S TRIANGLE

and the method of extracting roots depends upon the expressing of a number in the above form.

An approximate value of the cube root of 998 may be found by this means:

$$998 = 1000 - 2$$
$$= 1000(1 - \cdot 002)$$

therefore
$$\sqrt[3]{998} = \sqrt[3]{1000(1 - \cdot 002)}$$
$$= (10)\,(\sqrt[3]{1 - \cdot 002})$$
$$= 10(1 - \cdot 002)^{\frac{1}{3}}$$

We now expand the part of the expression $(1 - \cdot 002)^{\frac{1}{3}}$ where $x = -\cdot 002$; and $n = \frac{1}{3}$. The expression $(1 - \cdot 002)^{\frac{1}{3}}$, by the modified expansion, is approximately equivalent to:

$$1 - \tfrac{1}{3}(\cdot 002)$$
$$= 1 - \cdot 007$$
$$= \cdot 9993$$

therefore
$$\sqrt[3]{998} = 10 \times \cdot 9993$$
$$= 9 \cdot 993 \text{ (approx.)}$$

Numbers may, of course, be raised to higher powers by the same method:

$$998^2 = 1000^2(1 - \cdot 002)^2$$
$$= 1000^2(1 - 2 \times \cdot 002)$$
$$= 1000^2(\cdot 996)$$
$$= 996{,}000$$

The degree of error is remarkably small, for the actual value of 998^2 is 996,004.

The principle of ignoring terms of an expansion if they are too small to be of importance may also be employed to evaluate expressions of the form:

$$(1 + x)\,(1 + y)$$

such as
$$(1 \cdot 002)\,(1 \cdot 003)$$

for where both x and y are very small, their product xy will be very much smaller still, so that:

$$(1 + x)\,(1 + y) = 1 + x + y \text{ approx.}$$

Thus $(1 \cdot 002)(1 \cdot 003) = 1 + \cdot 002 + \cdot 003$ approx.
$$= 1 \cdot 005$$

The degree of error is again very small, the actual value being
$$1 \cdot 005006$$

By similar methods, the expression:
$$\frac{(1+x)}{(1+y)}$$
may be reduced to $\quad 1 + x - y$

where x and y are very small.

Thus,
$$\frac{1 \cdot 005}{1 \cdot 002} = 1 + \cdot 005 - \cdot 002$$
$$= 1 \cdot 003$$

APPENDIX 2

Triangular Numbers and Combinations

The identity ^nC_x is used to express the number of different combinations which may be made of n things taken x at a time, and is calculated according to the following expansion:

$$\frac{n(n-1)\ (n-2)\ (n-3)}{(x-1)\ (x-2)\ (x-3)} \cdot \cdot \cdot$$

The number of factors in both denominator and numerator is equivalent to x. Thus, the number of combinations of 9 things taken 3 at a time (that is, $n=9$; $x=3$)

$$=\frac{9\times 8\times 7}{1\times 2\times 3}=84$$

The expansion ^nC_x displays a relationship between itself and the triangular numbers referred to in Chapter 2. This relationship arises out of the fact that the number of combinations of any number of different things taken *two* at a time will always be a triangular number. The expansions for different values of n are as follows:

$n=$	2	3	4	5	. . .	10
Expansion:	$^2C_2$	$^3C_2$	$^4C_2$	$^5C_2$	. . .	$^{10}C_2$
Combinations:	1	3	6	10	. . .	45

We have previously noted that triangular numbers are equal to the sums of series of consecutive integers from 1 upwards. Thus:

$$1+2 = 3$$
$$1+2+3 = 6$$
$$1+2+3+4=10$$

where 3, 6 and 10 and all similar numbers are triangular.

If we treat the triangular numbers as terms of series and find the sums of these series, we have:

$$
\begin{aligned}
1+3 &= 4 \\
1+3+6 &= 10 \\
1+3+6+10 &= 20 \\
1+3+6+10+15 &= 35
\end{aligned}
$$

From this we derive the new series 1, 4, 10, 20, 35, etc., and these numbers are found to be equal to the number of combinations of things taken *three* at a time:

$$
\begin{array}{ccccc}
^3C_3 & ^4C_3 & ^5C_3 & ^6C_3 & ^7C_3 \\
= 1 & 4 & 10 & 20 & 35
\end{array}
$$

From these facts it is clear that $^nC_3$, for example, may be expressed as an addition of terms of the form $^nC_2$. Thus:

$$^4C_3 = {}^2C_2 + {}^3C_2$$

and similarly $\quad ^7C_3 = {}^2C_2 + {}^3C_2 + {}^4C_2 + {}^5C_2 + {}^6C_2$

Numerically, the latter example is expressed as:

$$
\begin{aligned}
\frac{7 \cdot 6 \cdot 5}{1 \cdot 2 \cdot 3} &= \frac{2 \cdot 1}{1 \cdot 2} + \frac{3 \cdot 2}{1 \cdot 2} + \frac{4 \cdot 3}{1 \cdot 2} + \frac{5 \cdot 4}{1 \cdot 2} + \frac{6 \cdot 5}{1 \cdot 2} \\
&= \frac{(2 \cdot 1) + (3 \cdot 2) + (4 \cdot 3) + (5 \cdot 4) + (6 \cdot 5)}{2}
\end{aligned}
$$

an expression which, despite its obvious pattern, would be difficult to prove (short of calculating each side in full) without the use of the combination formula allied to the theory of the nature of triangular numbers.

APPENDIX 3

The Four Fours Problem—Examples

This very old problem consists of expressing successive integers (up to a specified limit) in the appropriate mathematical form, using only four fours in each expression together with any necessary signs. This throws an interesting light upon the construction of numbers and a few examples are therefore given below:

$$1 = \frac{4}{4} \times \frac{4}{4}$$

$$2 = \frac{4}{4} + \frac{4}{4}$$

$$3 = \frac{4+4+4}{4}$$

$$4 = \frac{4}{\sqrt{4}} \times \frac{4}{\sqrt{4}}$$

$$5 = \sqrt{4} + \sqrt{4} + \frac{4}{4}$$

$$6 = (\sqrt{4})\left(4 - \frac{4}{4}\right)$$

$$7 = \frac{44}{4} - 4$$

$$8 = \frac{4 \times 4}{4} + 4$$

$$9 = 4 + 4 + \frac{4}{4}$$

$$13 = \sqrt{4} + \sqrt{4} + \frac{4}{.4}$$

$$14 = 4 \times 4 - \frac{4}{\sqrt{4}}$$

$$15 = 4 \times 4 - \frac{4}{4}$$

$$16 = 4 + 4 + 4 + 4$$

$$17 = 4 \times 4 + \frac{4}{4}$$

$$18 = 4 \times 4 + \frac{4}{\sqrt{4}}$$

$$19 = \frac{4}{.4} + \frac{4}{.4}$$

$$20 = 4\left(4 + \frac{4}{4}\right)$$

$$21 = \frac{44 - \sqrt{4}}{\sqrt{4}}$$

$$10 = \frac{44-4}{4} \qquad 22 = \frac{4+4}{\cdot 4} + \sqrt{4}$$

$$11 = \frac{4}{\cdot 4} + \frac{4}{4} \qquad 23 = \frac{44 + \sqrt{4}}{\sqrt{4}}$$

$$12 = \frac{44+4}{4} \qquad 24 = \frac{44}{\sqrt{4}} + \sqrt{4}$$

A related problem is to represent the number 100 using only four identical digits. Thus,

$$100 = (5+5)(5+5)$$

or
$$= 99\tfrac{9}{9}$$

or
$$= \text{Antilog}(\tfrac{1}{1} + \tfrac{1}{1})$$

or
$$= \frac{7}{\cdot 7} \times \frac{7}{\cdot 7}$$

or
$$= (4 + 4 + \sqrt{4})^{(\sqrt{4})}$$

The third and fourth examples will be found to apply for any digit, and not only for those shown.

APPENDIX 4

'Naming the Day'

The twin facts that every week has seven days and that every year (other than leap years) has a fixed cycle of 365 days enable us, by making allowance for the leap years intervening, to relate any two days in time by a set formula. Examples of the fixed relationship between the days of any two weeks may be seen on any calendar in any month, thus:

M.	T.	W.	T.	F.	S.	S.
1	2	3	4	5	6	7
8	9	10	11	12	13	14
15	16	17	18	19	20	21
22	23	24	25	26	27	28
29	30	31	—	—	—	

If a rectangle be drawn anywhere on the calendar, so as to enclose some of the numbers but excluding the blank spaces, then the numbers at the opposite ends of one diagonal when added together will be equivalent to the total of the two end numbers on the other diagonal. Thus, from the above example:

$$5+27 = 6+26$$
and $$15+31 = 17+29$$

If the rectangle is a square, then the same total is also provided by the top and bottom numbers in the centre column and also by the first and last numbers in the centre row.

Thus,
$$15+31 = 17+29$$
$$= 16+30$$
$$= 22+24$$

A square of this type is an incomplete magic square and the reasons why such a pattern results is simply explained. Every

Wednesday, for example, is 2 days later than the preceding Monday. Every Monday is 7 days later than the preceding Monday. If the date of the first Monday in any month be the 1st, then other dates are calculated:

	M.	T.	W.
	1	2	3
	8	9	10

or

	M.	T.	W.
	A 1	1+1	1+2 B
	C (1+7)	(1+7)+1	(1+7)+2 D

and it is obvious that

$$A+D=B+C$$

Just as there is a fixed relationship between the first day and any other day of a month, so there is also a similar though more complex relationship between the first day and any other day of a complete century. Also, in the same way as we can calculate the date of any day, so may we perform the reverse operation of ascertaining on what day of the week any date will fall.

If we confine ourselves to the twentieth century beginning 1900 we may proceed as follows for any date. Add together:

(a) The last two digits of the year.
(b) A quarter of this same figure (ignoring fractions).
(c) The code number for the month (see below).
(d) The number of the day in the month.

The code numbers for the months January to December are, in order,

0, 3, 3, 6, 1, 4, 6, 2, 5, 0, 3, 5

Therefore, to find the day on which the 25th December 1954 fell, we have:

(a) 54
(b) 13
(c) 5
(d) 25
___
    97

This number is then divided by 7 and the remainder from this division indicates the position of the day in the week. If there is no remainder, the required date is a Sunday.

$$97 \div 7 = 13 \text{ (remainder 6)}$$

and the required date therefore falls on the sixth day after Sunday; that is, on a Saturday.

For the century 1800–1899, the procedure is the same but the monthly code numbers are different. They are, in order:

2, 5, 5, 1, 3, 6, 1, 4, 0, 2, 5, 0

Thus for the date 25th December 1854, we have:

(a) 54
(b) 13
(c) 0
(d) 25
─────
92 ÷ 7 = 13 (remainder 1)

so that this date fell on a Monday.

In point of fact, however, there is no need to memorize the different monthly code numbers for the century 1800–1899. If it is desired to find the day on which any date fell in that century, this may be found just as easily by working out the same date in the twentieth century and *adding 2 to the total*. The total for the date 25.12.1954 was 97. Adding 2 to this total gives 99. The division of the latter number by 7 leaves a remainder of 1, which is the same as the remainder resulting from the division of the total (92) for the date 25.12.1854.

It is not known who first discovered this formula. It is mentioned in Walsh's *Handbook of Literary Curiosities*, published in 1892 (though the monthly code numbers quoted there are incorrect), but its origin is probably much older.

Another interesting fact is that in any year other than a leap year, the first and last days of the year fall on the same week-day in their respective weeks. This is because the last day is 364 days later than the first day, and the number 364 is exactly divisible by 7.

APPENDIX 5

Series and the Weights Problems

A mathematical problem of great antiquity is that wherein it is required to find the least number of different weights which would enable us to weigh any integral number of pounds, from 1 pound to a specified upper limit. The usual problem has an upper limit of 40 lb.

It is first to be noted that there are two different ways of using the weights and the problems usually stipulate either (a) that the weights may be placed in only one of the scale-pans, or (b) that they may be placed in *either* of the pans so that at any one time there may be one or more weights in one or both scale-pans. There are therefore really two problems, but both solutions are in the form of geometric progressions.

If only one pan may contain weights then, as stated by Tartaglia in 1556, the weights required are:

$$1 \quad 2 \quad 4 \quad 8 \quad 16 \quad 32$$

whereas if both pans may be used simultaneously, the weights required are:

$$1 \quad 3 \quad 9 \quad 27$$

By the second method the weight of 2 lb., for instance, is obtained by placing the 1-lb. and 3-lb. weights in opposite pans, and the same principle is repeated where necessary for other weights. This solution is due to Bachet, who published it in 1624.

Bachet could not prove that his solution was the only one possible with all-different weights or that it used the least number of weights, but this was subsequently proved in 1886 by Major MacMahon.

SERIES AND WEIGHTS PROBLEMS

In general, it is found that, if the total number of pounds we wish to take as the upper limit is expressed in the form $\frac{1}{2}(3^n - 1)$, then the weights required to weigh any amount up to that limit will be:

$$1 \quad 3 \quad 3^2 \quad 3^3 \quad \ldots \quad 3^{n-1}$$

If 40 be expressed in this form, then:

$$3^n = 81 = 3^4$$

and the solution is therefore:

$$1 \quad 3 \quad 9 \quad 27$$
or
$$1 \quad 3^1 \quad 3^2 \quad 3^3$$

This solution depends upon the facts that every positive integral number can be expressed in terms of positive powers of the number 3 (if we employ minus signs). Similarly, the solution:

$$1 \quad 2 \quad 4 \quad 8 \quad 16 \quad 32$$

depends upon the fact that all numbers can be expressed in terms of the number 2 without the employment of minus signs.

There is no other number which satisfies the conditions that all other numbers may be expressed in terms of its positive integral powers, and there cannot, therefore, be any further solutions to the weights problems, *provided all the weights must be different.*

Index

Amicable numbers, 53, 90
Apocalyptic Number, 158
Archimedes, 114
Automorphic numbers, 90

Bachet, 172
Beast, Number of, 158, 160
Binomial theorem, 31, 161

Calculating prodigies, 133-4
Calendar, continuous, 169
Combinations, 165
Composite numbers, 55-63
Congruences, 108-11
Continued fractions, 114-22
'Counting cards', 132
Counting methods, 9
Cubes, 17, 27-30, 33-7, 43

Decimals, 15, 16
Decimals, cycles, 15, 50, 94-107
Digital roots, 38, 98, 131
Digits, Grouping of, 10, 126
Digit symbols, 11
Divisibility, 48, 64-71, 109

e, 121-3
Eratosthenes, Sieve of, 46
Euclid, 49, 52, 87, 112
Euler, 49, 52, 54, 121-3

Factorial numbers, 31
Factors, 46-63, 87-90
Fallacies, 135-44

Fermat, 37, 47, 54
Fibonacci numbers, 118
Fractional numbers, 14, 15
Fractions, continued, 114-22

Gauss, 108

Hexagonal numbers, 28, 29, 44

Imaginary numbers, 112, 122-124
Infinitude of numbers, 15
Infinity, 13
Integral numbers, 14, 15
Irrationals, 15, 112

Logarithms, 81-6, 139

MacMahon, Major, 172
Magic squares, 145-55
Mersenne's numbers, 54
Modulus, 108
Multiplication, 72-80

Napier, 83
Natural numbers, 63
Negative numbers, 12, 14, 15
Notation, scales of, 11, 159
Nought, 12
Number mystery, 156-60
Number origins, 9
Numerology, 159-60

Oblong numbers, 26

Palindromic numbers, 90, 91
Pascal's triangle, 161
Percentages, 143
Perfect numbers, 87–9, 158
Pi, 112–22, 124
Positive numbers, 12, 14, 15
'Power' numbers, 30–4
Primes, 46–63, 111
Probability, 31
Pythagorean numbers, 22–4, 83

Rational numbers, 15
Residues, 108
Russell, A. H., 67

Sarrus, 55

Scales of Notation, 11, 159
Series, 17–37, 81, 89, 102, 139, 146, 172
Series, consecutive odd numbers, 21, 24, 25, 27–30, 48, 55–63
Squares, 17–27, 29–35, 37, 43
Statistics, 142

Tartaglia, 172
Triangular numbers, 17–21, 28, 44, 89, 165
Trigonometrical ratios, 81–6

Weights problems, 172
Wilson's theorem, 111